Short-range Wireless Communication

Fundamentals of RF System Design and Application

Short-range Wireless Communication

Fundamentals of RF System Design and Application
Second Edition

by Alan Bensky

AMSTERDAM • BOSTON • HEIDELBERG • LONDON
NEW YORK • OXFORD • PARIS • SAN DIEGO
SAN FRANCISCO • SINGAPORE • SYDNEY • TOKYO

ELSEVIER

Newnes

Newnes is an imprint of Elsevier
200 Wheeler Road, Burlington, MA 01803, USA
Linacre House, Jordan Hill, Oxford OX2 8DP, UK

Library of Congress Cataloging-in-Publication Data

(Application submitted.)

British Library Cataloguing-in-Publication Data
A catalogue record for this book is available from the British Library.

ISBN: 978-0-7506-7782-0

For information on all Newnes publications
visit our website at www.newnespress.com

Transferred to Digital Printing in 2011

Contents

Contents

Dedication

To my wife Nuki,
and to daughters Chani, Racheli, and Ortal

Preface to the First Edition

Developers, manufacturers and marketers of products incorporating short-range radio systems are experts in their fields—security, telemetry, medical care, to name a few. Often they add a wireless interface just to eliminate wires on an existing wired product. They may adapt a wireless subsystem, which is easy to integrate electrically into their system, only to find that the range is far short of what they expected, there are frequent false alarms, or it doesn't work at all. It is for these adapters of wireless subsystems that this book is primarily intended.

Other potential readers are curious persons with varied technical backgrounds who see the growing applications for wireless communication and want to know how radio works, without delving deeply into a particular system or device. This book covers practically all aspects of radio communication including wave propagation, antennas, transmitters, receivers, design principles, telecommunication regulations and information theory. Armed with knowledge of the material in this book, the reader can more easily learn the details of specialized radio communication topics, such as cellular radio, personal communication systems (PCS), and wireless local area networks (WLAN).

The technical level of this book is suitable for readers with an engineering education or a scientific background, working as designers, engineering managers, or technical marketing people. They should be familiar with electrical circuits and engineering mathematics. Elementary probability theory is needed in some of the early chapters. Readers without an appropriate background or who need to brush up on probability are advised to jump ahead to chapter 10.

The book is organized as follows:

Chapter 1 is an introduction, presenting the focus of the book and the types of short-range radio applications that are covered.

Chapter 2 discusses radio propagation and factors that affect communication range and reliability.

Chapter 3 reviews the antennas used in short-range radio as well as transmission lines and circuit-matching techniques.

Chapter 4 covers the various forms of signals used for information transmission and modulation, and overall wireless system properties.

Chapters 5 and 6 describe the various kinds of transmitters and receivers.

Chapter 7 details the performance characteristics of radio systems.

Chapter 8 presents various component types that can be used to implement a short-range radio system.

Chapter 9 covers regulations and standards. It gives an overview of the conditions for getting approval of short-range radio systems in North America and Europe.

Chapter 10 is an introduction to probability and communication theory.

Chapter 11 reviews some of the most important new developments in short-range radio.

An introductory section describes the twelve Mathcad worksheets included on the companion website accompanying the book, which are helpful for wireless design engineers in their daily work. A fully searchable pdf version of the book is also included on the companion website.

Several terms in the book are used synonymously for varied expression, although there are subtle differences. "Wireless" and "radio" are used without distinction, although generally "wireless" also includes infrared communication and power line communication, which are not covered in this book. "Short-range radio" and "low-power radio" both refer to the area of unlicensed radio communication, although low power can be used to communicate over thousands of kilometers whereas short range, as used here, refers to several kilometers at the very most.

The book has a number of schematic diagrams, most of which do not include component values. Circuit design is more involved than just copying values from a schematic, and my intent is to explain concepts and give initial direction to engineers who have the ability to design a circuit to their own specific requirements. Many "cookbook" texts are available to assist in the actual circuit development as needed.

I wish to thank Professor Moe Bergman, who encouraged and assisted me from the time of my early interest in radio communication, for reviewing the manuscript and offering many helpful suggestions.

Preface to the Second Edition

Deployment of short-range wireless devices has grown steadily since the appearance of garage door openers and other keyless entry devices, but there as been no parallel to the increase in quantities of products in this category that were produced during the three years since the first edition of this book was published. WLAN has solidified its acceptance in the workplace, not only complimenting wired LAN but often displacing it entirely. While it has been expected for years that WLAN would gain a foothold in multicomputer homes, the distinction between corporate and home WLAN requirements has all but been erased, and during this period we have seen the Wi-Fi standard becoming ubiquitous, pushing out of the way the HomeRF system that was developed specifically for residential use. Another phenomenon, both nourished by and encouraging the inclusion of Wi-Fi in portable computers, is the spread of hot spots in the U.S.A., Europe, and other regions. Through these public access points, the growth of wireless networks is being accelerated by the desire for internet connections, anytime, anywhere.

Another networking industrial standard, Bluetooth, is rapidly gaining acceptance. Delayed somewhat in comparison to early expectations, sales of Bluetooth chips are rising fast, and the devices are finding their way into more and more cell phone models and associated wireless headphones, wireless USB adapters, and laptop, notebook and hand held computers.

Along with the greatly increased density of short-range wireless transmissions, with virtually no expansion of available frequencies in the unlicensed bands, there is naturally greater pressure to adapt newer technologies to permit higher spectrum utilization. One of these is ultra-wide band. Generated at baseband and having a broad noise-like spectrum from which information can be detected, UWB has been used up to now in predominantly military applications, such as virtually undetectable communications, and ground and wall penetrating radar. It is necessary for the regulatory authorities to redefine the way they insure coexistence between the myriad users of radio communication in order for UWB to get a commercial foothold. The FCC did this in 2002, with an addition to its regulations relating to unlicensed UWB transmissions, and several companies are meeting the challenge by developing revolutionary products with breakthroughs in data rates and distance measurement capability.

Inevitably, the requirements for very high functional density, high production quantities, and low prices that encourage mass acceptance are fulfilled in large part by changes and improvements in basic components. Not only the components used directly in the industry standard devices like Wi-Fi and Bluetooth are affected. Similar technologies are adopted to raise performance and lower prices in IC's and other components used for proprietary devices as well. Many of the components that were the heart of wireless devices three years ago are no longer available. Prominent companies that produced lines of integrated circuits for unlicensed band devices have left the field or are concentrating on the mass consumer products like cellular, Wi-Fi, and Bluetooth, relinquishing the task of providing IC's to the alarm, control and short range telemetry product designers to other, often smaller firms, who have accepted the challenge in vigor.

Considering all said above, a second edition became mandatory for a book that aims to be an active development tool as well as up-to-date reference for anyone designing short-range wireless devices or integrating them into electronic products. These are the principle changes in the book:

- Chapter 8, System Implementation, has been revamped. Discontinued devices were removed, and there are short descriptions of a wide range of devices from many manufacturers, which demonstrate the scope of components available and provide a starting point for developers who need to choose from various options to meet their requirements.

- Chapter 9, Regulations and Standards, was updated to include important changes to FCC regulations pertaining to unlicensed systems. This chapter now includes the definitions for the U-NII bands, used by the 802.11a version of Wi-Fi. Regulatory limits for communications applications of UWB are detailed in this edition. Chapter 9 also includes a review of the basics of the R&TTE Directive, important for anyone who intends to market wireless communication devices in Europe. An important addition to Chapter 9 is the inclusion of technical requirements pertaining to the certification of short-range/low power wireless devices in Japan.

■ Chapter 11, renamed Applications and Technologies, has been significantly expanded. Wi-Fi, Bluetooth and the new Zigbee network are described in considerable detail. The section on coexistence and compatibility now presents a criterion for estimating interference between Wi-Fi and Bluetooth transmissions and describes the methods being proposed to improve coexistence between networks of different standards operating on the same frequency band. In tune with the increased importance of UWB, now that it has been included in the FCC Rules, a more thorough description of its operation is provided, along with graphic presentations that make the explanation easier to understand.

In addition to these major changes, corrections and minor modifications were made in several other chapters of the book. There are additions to the References and Bibliography, including Web site listings. Finally, three Mathcad worksheets were added and existing ones updated or corrected.

I have no illusions that an engineering book dealing with a rapidly evolving subject such as short-range wireless can remain completely up-to-date for many years after its publication. However, the technology updates in this edition, among them new modulation methods such as CCK and OFDM, and the exposition of the up-and-coming UWB technology, should provide the insights a reader requires to understand new and related developments as they come across his path. While I've tried to give a broad but thorough treatment of short-range wireless communication as we see it now, I hope the second edition will serve as a key to understanding advances and new technologies that will inevitably continue to appear, some of them due to the work of readers of this book.

Alan Bensky, August 2003

What's on the Companion Website
Radio Engineering Worksheets

Included on the companion website accompanying this book are fifteen radio engineering worksheets. Throughout the text, sections that have an accompanying worksheet are indicated by this icon: 📖 . These worksheets will help you solve a wide variety of problems and should be of assistance to you in radio system design. The worksheets are based on Mathcad, a popular mathematics program published by MathSoft in which formulas and data are entered in familiar mathematical format, just as they would be when solving problems using pencil and paper or writing on the blackboard.

In order to use the worksheets you have to have Mathcad 2000 Professional or higher installed on your computer. The companion website contains a full performance evaluation version of Mathcad 11. The program remains valid for 120 days after installation. Use beyond that time is contingent upon purchasing a license from *Mathsoft* and following the activation procedure that comes with the software. You may want to print out the worksheets during the evaluation period in order to refer to them later if you do not opt to obtain the permanent version immediately when the evaluation copy expires.

To install the evaluation version of Mathcad 11, access the companion website and open the "AcademicCD" folder and activate "Setup.exe." Click "Mathcad 11," check "Evaluation Copy," then go through the install procedure as displayed. When it has completed, open "Mathcad 11" from "Start" >> "Programs." Check "Activate Later" on the screen that is presented and press "Finish." It is recommended to go through the tutorials if you are not familiar with Mathcad. Then open the worksheet you want to use from the "Mathcad Worksheets" folder on the companion website."

The Mathcad web site, www.mathcad.com, has a library of engineering worksheets that you can access and work on using the Mathcad 11 evaluation program.

Using the Worksheets

As stated above, Mathcad formulas appear in normal written form on the worksheets. There are some small differences in interpretation of symbols.

For example, a data entry expression is made up of a variable on the left, followed by a special equal sign which looks like a colon and an equals sign as follows := . An equals sign alone is followed by a calculated answer. It's worthwhile to study the HELP contents in order to benefit most from the worksheets, as well as to do your own mathematical calculations.

Text, interspersed with the mathematics, explains the organization of the worksheet and tells you where to enter data and where the answers are. All worksheets have default data that you replace with your data to solve specific problems. Note the following:

- Yellow marked expressions on the worksheet indicate where to insert your data. Click the cursor on the default data and erase it using the delete and backspace keys. Type in your numerical data on the remaining small black rectangle.

- Blue marked expressions are the calculated answers. They change automatically when you change the data (see below).

Calculations are usually performed by the program automatically as soon as you change the data and press Enter. This can be annoying when you originally enter your data into the worksheet, so you can disable this feature by pressing "Math" on the upper bar and then "Automatic Calculation" to remove the check mark. When you finish entering data, press "Automatic Calculation" again. Now when you change your data, the answers and graphs will automatically update, as on a spreadsheet. You can also initiate calculation if it happens to be disabled by pressing F9.

Graphs

A couple of the worksheets have graphs. If you want to find a particular coordinate and the resolution of the axes is not sufficient, click on the graph with the right mouse button. Click on Trace. Move the cursor on the plot and see the coordinates in the Trace window.

Units of Measure

One of the special features of Mathcad is the ease of using units of measure. You don't have to use any conversion factors when changing units. For example, if the default unit of length in a yellow data input expression is cm (centimeters) and you prefer to enter your data in inches, simply insert the number of inches, then replace "cm" with "in." Similarly, the units of measure in the blue solution expressions can also be replaced.

Worksheet Descriptions

Each worksheet has a basic description and text to help you use it. More detailed descriptions are given in the following sections.

Charge Pump PLL.mcd — "Find Filter Constants for Charge Pump PLL"

The charge pump phase-locked loop is commonly used in the frequency determining block of transmitters and receivers. While the PLL circuitry is often integrated with other RF components, the filter components are usually external to the chip and must be determined by the system designer. The worksheet can be used to design second or third order filters. The latter provides higher attenuation of the reference oscillator spurs that appear in the spectrum of the phase locked output frequency.

Several parameters must be entered in order to find the filter component values. f_{ref} is the frequency that is applied to the phase comparator. Some PLL chips have a variable or fixed-reference divider and f_{ref} is the quotient of the input reference frequency and the internal divisor.

fp is the open-loop unity gain frequency, which should not be greater than one tenth of the reference frequency f_{ref}. Kφ is the output current of the pulses from the charge pump phase comparator, and is generally given in the specification of the PLL device. VFO sensitivity is Kv. It represents the slope of the frequency vs. control voltage of the VFO. If not specified, it can be measured by applying a variable control voltage to the VFO and measuring the change in output frequency per volt change of control voltage.

N is the division ratio of the divider that divides down the output frequency to the reference frequency. Loop damping is determined by φp. It should be set to 45 or 50 degrees. A higher value will cause over damping and slower response when changing frequencies, and a lower value will increase overshoot and may lead to instability.

In addition to computing the loop filter values, the worksheet has graphs showing open loop, closed loop, and filter responses, and the idealized time response of the PLL which may be observed on the VCO control voltage input.

Conversions.mcd — Impedance Transformations

This worksheet is intended for general use in circuit design. It is particularly helpful in designing impedance-matching networks, together with the worksheet "Matching.mcd." Sections (6) and (7) can be used in impedance matching when the source and load impedances are not pure resistances. In these cases, combine the reactance with the adjacent reactance of the matching network.

Diffraction.mcd — Diffraction

Here you can see one reason why radio reception is possible in places that don't have a line-of-sight view to the transmitter. Note that the diffraction phenomenon affects the signal strength in line-of-sight paths as well. This worksheet is more tutorial than practical, since its results are accurate only where there is only one barrier that has the shape of a knife edge. In most real situations there are several barriers of various shapes, and signal strength is also affected by reflections. However, it is interesting and informative to see the effects of changing the frequency on wave penetration into shadowed regions. The calculations for the plot are complicated and take time, so be patient!

Helical.mcd — Helical Antennas

Helical antennas are commonly used for portable short-range transmitters and receivers, and you can get a good start on the design of one using this worksheet. After you insert the global parameters — frequency, antenna diameter, and wire diameter—you have two choices for the remaining data. If you know the turns per inch of the winding for the antenna, start from section (1) and insert the data. The antenna height will then be calculated. In section (2) right click on the yellow expression for height, then click "Disable Evaluation" in the pop-up window. Check sections (3) through (7) to see the results of your design.

If both the height and the diameter of the antenna are known, enter the height in the yellow expression under (2). (If the expression had been disabled as shown above, there will be a small black rectangle in it. Right click on it and click "Enable Evaluation.") The required number of turns for resonance will be shown in the blue expression. Get more information from sections (3) through (7).

By changing the form factor of the antenna, you affect the radiation resistance and efficiency. Section (6) gives the total resistance that has to be matched.

The formulas in the worksheet assume a perfect ground plane, which is rarely the case for portable devices. So regard the results of the calculations as starting points in the design. Start with a few more turns than calculated, then trim the antenna until resonance, or maximum radiation or reception, is achieved.

Loop.mcd — Loop Antenna

The printed loop antenna is one of the most popular for small portable UHF transmitters. Enter your basic data — dimensions and frequency. Sections (1), (2) and (3) give results involving radiation efficiency. The result in (3) is most probably an understatement, since other factors not taken into account will reduce the efficiency. Among them are losses in the dielectric of the board and surrounding components and housing materials.

Section (4) of the worksheet gives an approximation of the loop inductance, helpful for matching to the transmitter output.

Matching.mcd — Impedance Matching

This worksheet presents several topologies for matching functional blocks of a radio circuit — antenna to receiver input, transmitter final stage to antenna, and matching between RF amplifier stages, for example. R1 is the resistance seen looking into the network when it is terminated by R2. If a resistance R1 is connected to the left side, then the resistance looking into the right side will be R2. Circuits (1) and (2) each have only one solution for a pair of values R1 and R2, whereas in circuits (3) and (4) the values of the matching components depend on the value chosen for Q.

The losses of the matching circuit components, particularly the inductors, should be taken into consideration when choosing the matching circuit configuration. An inductor can be represented as having a small resistance in series with it or a large resistance in parallel. These resistances can be manipulated to be part of the resistances being matched with the help of worksheet "Impedance Transformations." See the impedance matching examples in Chapter 3.

Microstrip.mcd — Microstrip Transmission Lines

This worksheet tells you the characteristic impedance of a printed circuit board conductor when the width is known, or the required conductor width to get a specified characteristic impedance. Note that the conductor has to be backed by a ground plane on the opposite side. You have to specify the frequency, dielectric constant of the board insulating material, and board thickness. The actual dielectric constant needed for determining characteristic impedance is a function of the board thickness and conductor width. In section (2) it is the width we are trying to find, so we need to follow an iterative process to get a true solution. That is why the width that is determined from the first trial is used as the "guess value" for a second run that results in a closer estimate of the true width for the required characteristic impedance.

The results include the wavelength in the board for the particular conductor width. This value is needed when using the Smith chart for designing printed circuit matching networks or reactive components.

Miscellaneous.mcd — "Miscellaneous"

Several useful calculations for RF engineers are performed by this worksheet. They are:

1) For given system impedance and power in dBm find power in watts (or mW) and volts rms, or knowing V_{rms} find dBm.

2) Find the number of turns for a single-layer air coil given core dimensions and inductance, or find inductance when dimensions and number of turns are known.

3) Find mismatch loss as a function of VSWR.

4) Find attenuation of the common RG-58C/U coax when length and frequency are known.

5) Calculate receiver sensitivity from noise figure, bandwidth, and prededection signal-to-noise ratio.

Noise Figure.mcd — "Noise Figure"

Knowing receiver composite noise figure is a prerequisite for calculating sensitivity. In the worksheet, stage noise figure and gain are input to matrices. The worksheet noise figure calculations can account for the location of the image rejection filter, or lack of it, in the receiver front end. A "modification flag" is entered into the second column of the noise fixture matrix according to the instructions in the document. Mixer noise figure should be single sideband. The worksheet can perform conversions between single and double-sideband noise figure values.

Patch.mcd — *Microstrip Patch Antennas*

You can design a square half-wave microstrip patch antenna using this worksheet. The formulas account for the "fringing distance," which makes the patch length somewhat smaller than a half wavelength in the board. A result of the worksheet is the impedance at the center of a board edge to which you have to match the input impedance of a receiver or output impedance of a transmitter. This matching is usually done by printed microstrip distributed components. If you want to match directly to a coaxial cable, solder the cable's shield to the groundplane on the opposite side of the board. Connect the center connector through a via insulated from the groundplane to the patch at the distance "x" from the center, which is displayed as a result in the last section of the worksheet.

Radiate.mcd — *Radio Propagation Formulas*

This worksheet is very handy for anyone dealing with wave propagation and antennas. Note that in section (1) a factor "L" is included, which contains circuit and system losses. This factor is not explicit in the other sections but such losses should be incorporated in the gain factors. For example, section (3) may be used to find the transmitter power needed to meet the FCC regulations for maximum field strength at a distance of 3 meters. When a loop antenna is used, the antenna losses may be 10 dB or more so G_t should be at the most 0.1 in this case.

Range.mcd — Open Field Range

Estimating the range of a low-power wireless communication system under open field conditions is much more relevant than using the free space equations. Using this worksheet you can see the nulls and the peaks of signal strength that are so often experienced in practice, and how they are influenced by operating frequency and antenna heights.

To use the worksheet you enter the operating frequency, transmitting and receiving antenna heights above ground, and polarity of the transmitting antenna. You also indicate the maximum distance for the plot. The calculations vectorily add the direct line-of-sight received signal strength and that of the signal reflected from the ground. "Normal" ground conductivity and permittivity are used, which you could change if you wish but usually it isn't necessary. It's assumed that the antennas have constant gain in elevation. The result is a plot of the isotropic path gain (the negative of the path loss) in decibels, representing the ratio of the power at the input of a receiver relative to the radiated transmitter power in the direction of the receiver with receiver antenna gain equal to zero dB. The open field path gain is calculated relative to the free space power at a reference distance, d_0, set to 3 meters. Then transmitter power is found from the free space power at d_0. Free space path gain is also plotted, for reference.

The last sections of the worksheet let you find, for a given transmitted power, the required receiver sensitivity for a given range in an open field, or the open field range when receiver sensitivity is known. Finding the sensitivity when power and range are known is straightforward, but to get the range from given power and sensitivity you have to find the abscissa of the curve (the distance) when the ordinate (path gain) is known. The worksheet has instructions for doing it.

Translines.mcd — Transmission Lines

This worksheet solves various useful formulas for working with transmission lines. First you enter the operating frequency and transmission line parameters. The calculations account for line loss, whose value you have to enter, but if you don't know it, you can enter 0 dB with little effect on short low-loss lines. If you do use line loss, you have to know it for the frequency of operation, as the loss increases with frequency. From the cable velocity factor and frequency, which you provide, the wavelength in the cable is calculated and displayed.

Section (1) gives relationships between often-used matching parameters and uses only the line characteristic impedance that was entered. Section (2) gives the transformation of the load impedance to the impedance seen looking into the line from the source, as well as the required load impedance for a specified impedance in the line at the source. You only have to enter the line length. Section (3) lets you find line lengths needed, for shorted or open lines, to get desired inductive or capacitive reactances.

Probabilities.mcd — Probability of Detection

It's very useful to be able to find the probability of errorless detection, or the probability of error, of sending a sequence of a particular number of bits over a transmission system. You can do this, when the probability of error of each bit is known, using this worksheet. You can also find from the worksheet how much an error-correcting code can improve (reduce) the probability of error. The calculations use the Hamming bound for the minimum number of code letters required to correct all errors up to a given number.

The worksheet states that because of the increase of error-correcting bits in the sent message, the time to send the information is increased when bit rate is constant. Increasing the bit rate to bring the information rate up to its value without error correction entails increased bandwidth and a consequential reduction of signal-to-noise ratio. The effect on bit error rate depends on the type of modulation used and, when the new probability of bit error is known, you can input it to the data sheet and get a better solution for the probability of error with error-correction coding.

The worksheet also calculates the probability of false alarms. False alarms occur when noise changes received bits so that a randomly received sequence is identical to an expected message when this message is not being sent.

A common way to improve the probability of detection of messages without using error-correction coding is to send the same message consecutively several times. You can see the degree of improvement of the probability of detection from the repetition of messages from the results displayed in the last section of the worksheet.

S-parameters.mcd — S-Parameters

Often manufacturers of active and passive components provide sets of *s*-parameters, which range over bands of operating frequencies and biasing currents and voltages. By plugging the relevant set of *s*-parameters into this worksheet, you can find the input and output impedances that you need to design the matching networks for these devices. The worksheet finds reflection coefficients and input and output impedances according to the corresponding load or source impedances applied to the components.

CHAPTER 1

Introduction

1.1 Historical Perspective

A limited number of short-range radio applications were in use in the 1970s. The garage door opener was one of them. An L-C tuned circuit oscillator transmitter and superregenerative receiver made up the system. It suffered from frequency drift and susceptibility to interference, which caused the door to open apparently at random, leaving the premises unprotected. Similar systems are still in use today, although radio technology has advanced tremendously. Even with greatly improved circuits and techniques, wireless replacements for wired applications—in security systems for example—still suffer from the belief that wireless is less reliable than wired and that cost differentials are too great to bring about the revolution that cellular radio has brought to telephone communication.

Few people will dispute the assertion that cellular radio is in a class with a small number of other technological advancements—including the proliferation of electric power in the late 19th century, mass production of the automobile, and the invention of the transistor—that have profoundly affected human lifestyle in the last century. Another development in electronic communication within the last 10 or so years has also impacted our society—satellite communication—and its impact is coming even closer to home with the spread of direct broadcast satellite television transmissions.

That wireless techniques have such an overwhelming reception is not at all surprising. After all, the wires really have no intrinsic use. They only tie us down and we would gladly do without them if we could still get reliable operation at an acceptable price. Cellular radio today is of lower quality, lower reliability, and higher price generally than wired telephone, but its acceptance by the public is nothing less than phenomenal. Imagine the consequences to lifestyle when electric power is able to be distributed without wires!

Considering the ever-increasing influence of wireless systems in society, this book was written to give a basic but comprehensive understanding of radio communication to a wide base of technically oriented people who either have a curiosity to know how wireless works, or who will contribute to expanding its uses. While most chapters of the book will be a gateway, or even a prerequisite, to understanding the basics of all forms of radio communication, including satellite and cellular systems, the emphasis and implementations are aimed at what are generally defined as short-range or low-power wireless applications. These applications are undergoing a fast rate of expansion, in large part due to the technological fall-out of the cellular radio revolution.

1.2 Reasons for the Spread of Wireless Applications

One might think that there would be a limit to the spread of wireless applications and the increase in their use, since the radio spectrum is a fixed entity and it tends to be depleted as more and more use is made of it. In addition, price and size limitations should restrict proliferation of wire replacement devices. However, technological developments defy these axioms.

- We now can employ higher and higher frequencies in the spectrum whose use was previously impossible or very expensive. In particular, solid-state devices have recently been developed to amplify at millimeter wavelengths, or tens of gigahertz. Efficient, compact antennas are also available, such as planar antennas, which are often used in short-range devices. The development of surface acoustic wave (SAW) frequency-determining components allow generation of UHF frequencies with very simple circuits.

- Digital modulation techniques are replacing the analog methods of previous years, permitting a multiplication of the number of communication channels that can occupy a given bandwidth.

■ We have seen much progress in circuit miniaturization. Hybrid integrated circuits, combining analog and digital functions on one chip, and radio-frequency integrated circuits are to a large part responsible for the amazingly compact size of cellular telephone handsets. This miniaturization is not only a question of convenience, but also a necessity for efficient design of very short-wavelength circuits.

1.3 Characteristics of Short-range Radio

"Short-range" and "low-power" are both relative terms, and their scope must be asserted in order to see the focus of this book. Hardly any of the applications that we will discuss will have all of these characteristics, but all of them will have some of the following features:

■ RF power output of several microwatts up to 100 milliwatts

■ Communication range of several centimeters up to several hundred meters

■ Principally indoor operation

■ Omnidirectional, built-in antennas

■ Simple construction and relatively low price in the range of consumer appliances

■ Unlicensed operation

■ Noncritical bandwidth specifications

■ UHF operation

■ Battery-operated transmitter or receiver

Our focus on implementation excludes cellular radios and wireless telephones, although an understanding of the material in this book will give the reader greater comprehension of the principles of operation of those ubiquitous devices.

Short-range radio applications

The following table lists some short-range radio applications and characteristics that show the focus of this text.

Application	*Frequencies (MHz)*	*Characteristics*
Security Systems	300-500, 800, 900	Simplicity, easy installation
Emergency Medical Alarms	300-500, 800	Convenient carrying, long battery life, reliable
Computer Accessories— mouse, keyboard	UHF	High data rates, very short range, low cost
RFID (Radio Frequency Identification)	100 KHz – 2.4 GHz	Very short range, active or passive transponder
WLAN (Wireless Local Area Network)	900 MHz, 2.4 GHz	High continuous data rates, spread-spectrum modulation, high price
Wireless microphones; Wireless Headphones	VHF, UHF	Analog high-fidelity voice modulation, moderate price
Keyless Entry—Gate openers	UHF	Miniature transmitter, special coding to prevent duplication
Wireless bar code readers	900 MHz, 2.4 GHz	Industrial use, spread spectrum, expensive

A new direction in short-range applications is appearing in the form of high-rate data communication devices for distances of several meters. This is being developed by the Bluetooth consortium of telecommunication and PC technology leaders for eliminating wiring between computers and peripherals, as well as wireless internet access through cellular phones. Mass production will eventually bring sophisticated communication technology to a price consumers can afford, and fallout from this development will surely reach many of the applications in the table above, improving their reliability and increasing their acceptance for replacing wiring.

1.4 Elements of Wireless Communication Systems

Figure 1-1 is a block diagram of a complete wireless system. Essentially all elements of this system will be described in detail in the later chapters of the book. A brief description of them is given below with special reference to short-range applications.

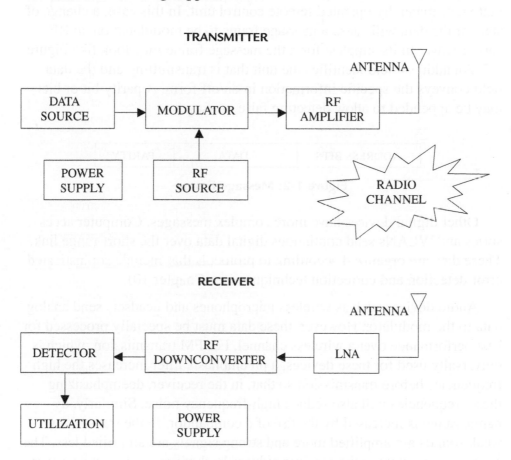

Figure 1-1: The Wireless System

Data source

This is the information to be conveyed from one side to the other. Each of the devices listed in the table on page 4 has its own characteristic data source, which may be analog or digital. In many of the cases the data may be simple on/off information, as in a security intrusion detector, panic button, or manually operated remote control unit. In this case, a change of state of the data will cause a message frame to be modulated on an RF carrier wave. In its simplest form the message frame may look like Figure 1-2. An address field identifies the unit that is transmitting and the data field conveys the specific information in on/off form. A parity bit or bits may be appended to allow detecting false messages.

ADDRESS BITS	DATA	PARITY

Figure 1-2: Message Frame

Other digital devices have more complex messages. Computer accessories and WLANs send continuous digital data over the short-range link. These data are organized according to protocols that include sophisticated error detection and correction techniques (see Chapter 10).

Audio devices such as wireless microphones and headsets send analog data to the modulator. However, these data must be specially processed for best performance over a wireless channel. For FM transmission, which is universally used for these devices, a preemphasis filter increases the high frequencies before transmission so that, in the receiver, deemphasizing these frequencies will also reduce high-frequency noise. Similarly, dynamic range is increased by the use of a compandor. In the transmitter weak sounds are amplified more and strong signals are amplified less. The opposite procedure in the receiver reduces background noise while returning the weak sounds to their proper relative level, thus improving the dynamic range.

A quite different aspect of the data source is the case for RFIDs. Here, the data are not available in the transmitter but are added to the RF signal in an intermediate receptor, called a transducer. See Figure 1-3. This transducer may be passive or active, but in any case the originally trans-

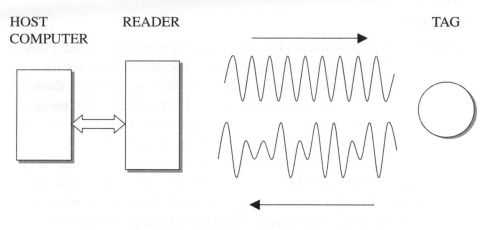

HOST COMPUTER   READER   TAG

Figure 1-3: RFID

mitted radio frequency is modified by the transducer and detected by a receiver that deciphers the data added and passes it to a host computer.

Radio frequency generating section

This part of the transmitter consists of an RF source (oscillator or synthesizer), a modulator, and an amplifier. In the simplest short-range devices, all three functions may be included in a circuit of only one transistor. Chapter 5 details some of the common configurations. Again RFIDs are different from the other applications in that the modulation is carried out remotely from the RF source.

RF conduction and radiation

Practically all short-range devices have built-in antennas, so their transmission lines are relatively short and simple. However, particularly on the higher frequencies, their lengths are a high enough percentage of wavelength to affect the transmission efficiency of the transmitter. Chapter 3 discusses the transmission lines encountered in short-range systems and the importance and techniques of proper matching. The antennas of short-range devices also distinguish them from other radio applications. They must be small—often a fraction of a wavelength—and omnidirectional for most uses.

Radio channel

By definition, the radio channel for short-range applications is short, and for a large part the equipment is used indoors. The allowed radio frequency power is relatively low and regulated by the telecommunication authorities. Also, the devices are often operated while close to or attached to a human (or animal) body, a fact which affects the communication performance. Reliable operating range is difficult to predict for these systems, and lack of knowledge of the special propagation characteristics of short-range radio by manufacturers, sellers, and users alike is a dominating reason for its reputation as being unreliable. Short-range devices are often used to replace hard wiring, so when similar performance is expected, the limitations of radio propagation compared to wires must be accounted for in each application. Chapter 2 brings this problem into perspective.

Receivers

Receivers have many similar blocks to transmitters, but their operation is reversed. They have an antenna and transmission line, RF amplifiers, and use oscillators in their operation. Weak signal signals intercepted by the antenna are amplified above the circuit noise by a low noise amplifier (LNA). The desired signal is separated from all the others and is shifted lower in frequency in a downconverter, where it may be more effectively amplified to the level required for demodulation, or detection. The detector fulfulls the ultimate purpose of the receiver; conversion of the data source which was implanted on the RF wave in the transmitter back to its original form.

While the transmitted power is limited by the authorities, receiver sensitivity is not, so the most obvious way to improve system performance is by improving the sensitivity and the selectivity to reduce interference from unwanted sources. This must be done under constraints of physics, cost, size, and often power consumption. Chapter 7 deals with these matters.

An important factor in low-power system design, and sometimes a controversial one, is the type of modulation to use. In the case of the simpler systems—security and medical alarms, for example—the choice is between amplitude shift keying (ASK), parallel to amplitude modulation in analog systems, and frequency shift keying (FSK), analogous to frequency modulation (FM). In Chapter 4 we'll look at the pros and cons of the two systems.

Power supplies

In most short-range devices, at least one side of the wireless link must be completely untethered—that's what wireless is for! When size is limited, as it is in hand-operated remote control transmitters and security detectors, battery size and therefore energy is limited. The need to change batteries often is not only highly inconvenient but also expensive, and this is an impediment to more widespread use of radio in place of wires. Thus, low-current consumption is an important design aim for wireless devices. This is usually harder to achieve for receivers than for transmitters. Many short-range applications call for intermittent transmitter operation, in security systems, for example. Transmitters can be kept in a very low-current standby status until data needs to be sent. The receiver, on the other hand, usually doesn't know when data will be sent so it must be alert all the time. Even so, there are techniques to reduce the receiver duty cycle so that it doesn't draw full current all the time. Another way to reduce receiver power consumption is to operate it in a reduced power standby mode, wherein operation goes to normal when the beginning of a signal is detected. This method often entails reduced sensitivity, however.

1.5 Summary

Short-range radio is developing as an expanding and distinct adjunct to wireless communication in general. While its basic operating characteristics are the same as all radio systems, there are many features and specific problems that justify dealing with it as a separate field. Among them are low power, low cost, small size, battery operation, uncertainty of indoor propagation, and unlicensed operation on crowded bands. The rest of this book will delve into the operational and design specialties of short-range radio communication from the electromagnetic propagation environment through antennas, receivers and transmitters, regulations and standards, and a bit of relevant information theory. The last chapter describes in detail current developments that are bringing wireless to the home at an unprecedented extent. Electronic worksheets contained on the accompanying CD-ROM and referred to throughout the book can be used to work out examples given in the text, and to help the reader solve his own specific design problems.

CHAPTER 2

Radio Propagation

It is fitting to begin a book about wireless communication with a look at the phenomena that lets us transfer information from one point to another without any physical medium—the propagation of radio waves. If you want to design an efficient radio communication system, even for operation over relatively short distances, you should understand the behavior of the wireless channel in the various surroundings where this communication is to take place. While the use of "brute force"—increasing transmission power—could overcome inordinate path losses, limitations imposed on design by required battery life, or by regulatory authorities, make it imperative to develop and deploy short-range radio systems using solutions that a knowledge of radio propagation can give.

The overall behavior of radio waves is described by Maxwell's equations. In 1873, the British physicist James Clerk Maxwell published his *Treatise on Electricity and Magnetism* in which he presented a set of equations that describe the nature of electromagnetic fields in terms of space and time. Appendix 2-A gives a brief description of those equations. Heinrich Rudolph Hertz performed experiments to confirm Maxwell's theory, which led to the development of wireless telegraph and radio. Maxwell's equations form the basis for describing the propagation of radio waves in space, as well as the nature of varying electric and magnetic fields in conducting and insulating materials, and the flow of waves in waveguides. From them, you can derive the skin effect equation and the electric and magnetic field relationships very close to antennas of all kinds. A number of computer programs on the market, based on the solution of Maxwell's equations, help in the design of antennas, anticipate electromagnetic radiation problems from circuit board layouts, calculate the effectiveness of shielding, and perform accurate simulation of ultra-high-frequency and microwave circuits. While you don't have to be an expert in Maxwell's equations to use these programs (you do in order to write them!), having some familiarity with the equations may take the mystery out of the operation of the software and give an appreciation for its range of application and limitations.

2.1 Mechanisms of Radio Wave Propagation

Radio waves can propagate from transmitter to receiver in four ways: through ground waves, sky waves, free space waves, and open field waves.

Ground waves exist only for vertical polarization, produced by vertical antennas, when the transmitting and receiving antennas are close to the surface of the earth (see *Polarization* under Section 3.2 in Chapter 3). The transmitted radiation induces currents in the earth, and the waves travel over the earth's surface, being attenuated according to the energy absorbed by the conducting earth. The reason that horizontal antennas are not effective for ground wave propagation is that the horizontal electric field that they create is short circuited by the earth. Ground wave propagation is dominant only at relatively low frequencies, up to a few MHz, so it needn't concern us here.

Sky wave propagation is dependent on reflection from the ionosphere, a region of rarified air high above the earth's surface that is ionized by sunlight (primarily ultraviolet radiation). The ionosphere is responsible for long-distance communication in the high-frequency bands between 3 and 30 MHz. It is very dependent on time of day, season, longitude on the earth, and the multi-year cyclic production of sunspots on the sun. It makes possible long-range communication using very low-power trans-mitters. Most short-range communication applications that we deal with in this book use VHF, UHF, and microwave bands, generally above 40 MHz. There are times when ionospheric reflection occurs at the low end of this range, and then sky wave propagation can be responsible for interference from signals originating hundreds of kilometers away. However, in general, sky wave propagation does not affect the short-range radio applications that we are interested in.

The most important propagation mechanism for short-range communication on the VHF and UHF bands is that which occurs in an open field, where the received signal is a vector sum of a direct line-of-sight signal and a signal from the same source that is reflected off the earth. We discuss below the relationship between signal strength and range in line-of-sight and open-field topographies.

The range of line-of-sight signals, when there are no reflections from the earth or ionosphere, is a function of the dispersion of the waves from the transmitter antenna. In this free-space case the signal strength decreases in inverse proportion to the distance away from the transmitter antenna. When the radiated power is known, the field strength is given by equation (2-1):

$$E = \frac{\sqrt{30 \cdot P_t \cdot G_t}}{d} \qquad (2\text{-}1)$$

where P_t is the transmitted power, G_t is the antenna gain, and d is the distance. When P_t is in watts and d is in meters, E is volts/meter.

To find the power at the receiver (P_r) when the power into the transmitter antenna is known, use (2-2):

$$P_r = \frac{P_t G_t G_r \lambda^2}{(4\pi d)^2} \qquad (2\text{-}2)$$

G_t and G_r are the transmitter and receiver antenna gains, and λ is the wavelength.

Range can be calculated on this basis at high UHF and microwave frequencies when high-gain antennas are used, located many wavelengths above the ground. Signal strength between the earth and a satellite, and between satellites, also follows the inverse distance law, but this case isn't in the category of short-range communication! At microwave frequencies, signal strength is also reduced by atmospheric absorption caused by water vapor and other gases that constitute the air.

2.2 Open Field Propagation

Although the formulas in the previous section are useful in some circumstances, the actual range of a VHF or UHF signal is affected by reflections from the ground and surrounding objects. The path lengths of the reflected signals differ from that of the line-of-sight signal, so the receiver sees a combined signal with components having different amplitudes and phases. The reflection causes a phase reversal. A reflected signal having a path length exceeding the line-of-sight distance by exactly the signal wavelength or a multiple of it will almost cancel completely the desired signal ("almost" because its amplitude will be slightly less than the direct signal amplitude). On the other hand, if the path length of the reflected signal differs exactly by an odd multiple of half the wavelength, the total signal will be strengthened by "almost" two times the free space direct signal.

In an open field with flat terrain there will be no reflections except the unavoidable one from the ground. It is instructive and useful to examine in depth the field strength versus distance in this case. The mathematical details are given in the Mathcad worksheet "Open Field Range."

In Figure 2-1 we see transmitter and receiver antennas separated by distance d and situated at heights h_1 and h_2. Using trigonometry, we can find the line of sight and reflected signal path lengths d_1 and d_2. Just as in optics, the angle of incidence equals the angle of reflection θ. We get the relative strength of the direct signal and reflected signal using the inverse path length relationship. If the ground were a perfect mirror, the relative reflected signal strength would exactly equal the inverse of d_2. In this case, the reflected signal phase would shift 180 degrees at the point of reflection. However, the ground is not a perfect reflector. Its characteristics as a reflector depend on its conductivity, permittivity, the polarization of the signal and its angle of incidence. In the Mathcad worksheet we have accounted for polarization, angle of incidence, and permittivity to find the reflection coefficient, which approaches –1 as the distance from the transmitter increases. The signals reaching the receiver are represented as complex numbers since they have both phase and amplitude. The phase is found by subtracting the largest interval of whole wavelength multiples from the total path length and multiplying the remaining fraction of a wavelength by 2π radians, or 360 degrees.

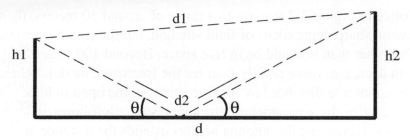

Figure 2-1: Open Field Signal Paths

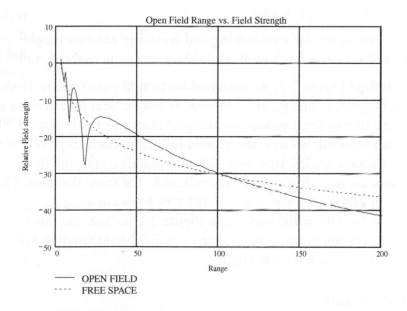

Figure 2-2: Field Strength vs. Range at 300 MHz

Figure 2-2 gives a plot of relative open field signal strength versus distance using the following parameters:

Polarity — horizontal
Frequency — 300 MHz
Antenna heights — both 3 meters
Relative ground permittivity — 15

Also shown is a plot of free space field strength versus distance (dotted line). In both plots, signal strength is referenced to the free space field strength at a range of 3 meters.

Notice in Figure 2-2 that, up to a range of around 50 meters, there are several sharp depressions of field strength, but the signal strength is mostly higher than it would be in free space. Beyond 100 meters, signal strength decreases more rapidly than for the free space model. Whereas there is an inverse distance law for free space, in the open field beyond 100 meters (for these parameters) the signal strength follows an inverse square law. Increasing the antenna heights extends the distance at which the inverse square law starts to take effect. This distance, d_m, can be approximated by

$$d_m = (12 \times h_1 \times h_2)/\lambda \tag{2-3}$$

where h_1 and h_2 are the transmitting and receiving antenna heights above ground and λ is the wavelength, all in the same units as the distance d_m.

In plotting Figure 2-2, we assumed horizontal polarization. Both antenna heights, h_1 and h_2, are 3 meters. When vertical polarization is used, the extreme local variations of signal strengths up to around 50 meters are reduced, because the ground reflection coefficient is less at larger reflection angles. However, for both polarizations, the inverse square law comes into effect at approximately the same distance. This distance in Figure 2-2 where λ is 1 meter is, from equation (2-3): $d_m = (12 \times 3 \times 3)/\lambda = 108$ meters. In Figure 2-2 we see that this is approximately the distance where the open-field field strength falls below the free-space field strength.

2.3 Diffraction

Diffraction is a propagation mechanism that permits wireless communication between points where there is no line-of-sight path due to obstacles that are opaque to radio waves. For example, diffraction makes it possible to communicate around the earth's curvature, as well as beyond hills and obstructions. It also fills in the spaces around obstacles when short-range radio is used inside buildings. Figure 2-3 is an illustration of diffraction geometries, showing an obstacle whose extremity has the shape of a knife edge. The obstacle should be seen as a half plane whose dimension is infinite into and out of the paper. The field strength at a receiving point relative to the free-space field strength without the obstacle is the diffraction gain. The phenomenon of diffraction is due to the fact that each point

Figure 2-3: Knife-edge Diffraction Geometry

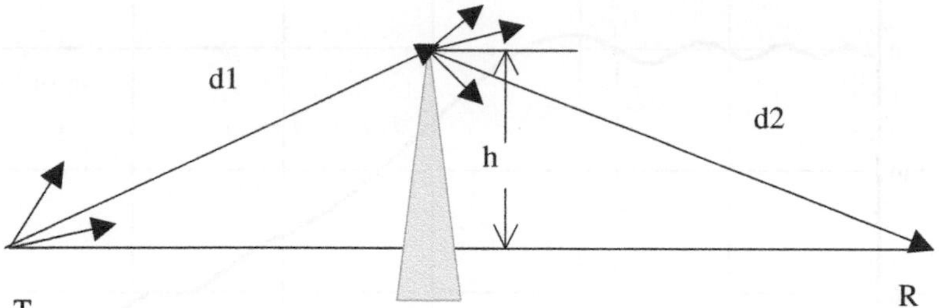

a) R is in shadow—h is positive

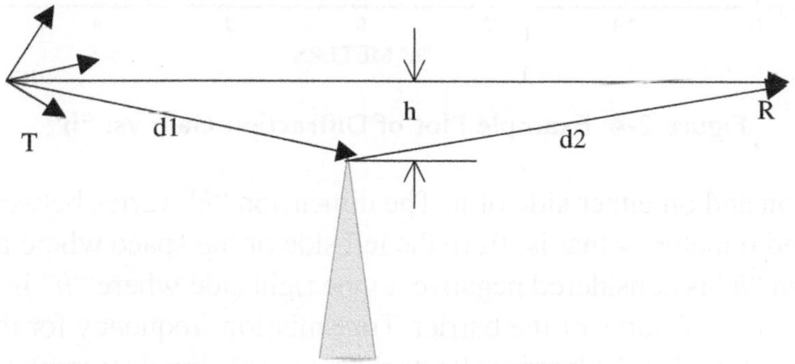

b) T and R are line of sight—h is negative

on the wave front emanating from the transmitter is a source of a secondary wave emission. Thus, at the knife edge of the obstacle, as shown in Figure 2-3a, there is radiation in all directions, including into the shadow.

The diffraction gain depends in a rather complicated way on a parameter that is a function of transmitter and receiver distances from the obstacle, d_1 and d_2, the obstacle dimension h, and the wavelength. Think of the effect of diffraction in an open space in a building where a wide metal barrier extending from floor to ceiling exists between the transmitter and the receiver. In our example, the space is 12 meters wide and the barrier is 6 meters wide, extending to the right side. When the transmitter and receiver locations are fixed on a line at a right angle to the barrier, the field strength at the receiver depends on the perpendicular distance from the line-of-sight path to the barrier's edge. Figure 2-4 is a plot of diffraction gain when transmitter and receiver are each 10 meters from the edge of the

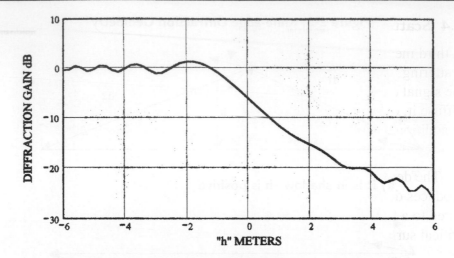

Figure 2-4: Example Plot of Diffraction Gain vs. "h"

obstruction and on either side of it. The dimension "*h*" varies between –6 meters and 6 meters—that is, from the left side of the space where the dimension "*h*" is considered negative, to the right side where "*h*" is positive and fully in the shadow of the barrier. Transmission frequency for the plot is 300 MHz. Note that the barrier affects the received signal strength even when there is a clear line of sight between the transmitter and receiver ("*h*" is negative as shown in Figure 2-3b). When the barrier edge is on the line of sight, diffraction gain is approximately –6 dB, and as the line-of-sight path gets farther from the barrier (to the left in this example), the signal strength varies in a cyclic manner around 0 dB gain. As the path from transmitter to receiver gets farther from the barrier edge into the shadow, the signal attenuation increases progressively.

Admittedly, the situation depicted in Figure 2-4 is idealistic, since it deals with only one barrier of very large extent. Normally there are several partitions and other obstacles near or between the line of sight path and a calculation of the diffraction gain would be very complicated, if not impossible. However, a knowledge of the idealistic behavior of the defraction gain and its dependence on distance and frequency can give qualitative insight. The Mathcad worksheet "Defraction" lets you see how the various parameters affect the defraction gain.

2.4 Scattering

A third mechanism affecting path loss, after reflection and diffraction, is scattering. Rough surfaces in the vicinity of the transmitter do not reflect the signal cleanly in the direction determined by the incident angle, but diffuse it, or scatter it in all directions. As a result, the receiver has access to additional radiation and path loss may be less than it would be from considering reflection and diffraction alone.

The degree of roughness of a surface and the amount of scattering it produces depends on the height of the protuberances on the surface compared to a function of the wavelength and the angle of incidence. The critical surface height h_c is given by [Gibson, p. 360]

$$h_c = \frac{\lambda}{8\cos\theta_i} \tag{2-4}$$

where λ is the wavelength and θ_i is the angle of incidence. It is the dividing line between smooth and rough surfaces when applied to the difference between the maximum and the minimum protuberances.

2.5 Path Loss

The numerical path loss is the ratio of the total radiated power from a transmitter antenna times the numerical gain of the antenna in the direction of the receiver to the power available at the receiver antenna. This is the ratio of the transmitter power delivered to a lossless antenna with numerical gain of 1 (0 dB) to that at the output of a 0 dB gain receiver antenna. Sometimes, for clarity, the ratio is called the *isotropic* path loss. An isotropic radiator is an ideal antenna that radiates equally in all directions and therefore has a gain of 0 dB. The inverse path loss ratio is sometimes more convenient to use. It is called the path gain and when expressed in decibels is a negative quantity. In free space, the isotropic path gain *PG* is derived from (2-2), resulting in

$$PG = \frac{\lambda^2}{(4\pi d)^2} \tag{2-5}$$

We have just examined several factors that affect the path loss of VHF-UHF signals—ground reflection, diffraction, and scattering. For a given site, it would be very difficult to calculate the path loss between transmitters and receivers, but empirical observations have allowed some general conclusions to be drawn for different physical environments. These conclusions involve determining the exponent, or range of exponents, for the distance d related to a short reference distance d_0. We then can write the path gain as dependent on the exponent n:

$$PG = k\left(\frac{d_0}{d}\right)^n \tag{2-6}$$

where k is equal to the path gain when $d = d_0$.

Table 2-1 shows path loss for different environments.

Table 2-1. Path Loss Exponents for Different Environments
[Gibson, p. 362]

Environment	Path gain exponent n
Free space	2
Open field (long distance)	4
Cellular radio — urban area	2.7 – 4
Shadowed urban cellular radio	5 – 6
In building line-of site	1.6 – 1.8
In building — obstructed	4 – 6

As an example of the use of the path gain exponent, let's assume the open field range of a security system transmitter and receiver is 300 meters. What range can we expect for their installation in a building?

Figure 2-5 shows curves of path gain versus distance for free-space propagation, open field propagation, and path gain with exponent 6, a worst case taken from Table 2-1 for "In building, obstructed." Transmitter and receiver heights are 2.5 meters, polarization is vertical, and the frequency is 915 MHz. The reference distance is 10 meters, and for all three curves the path gain at 10 meters is taken to be that of free space. For an

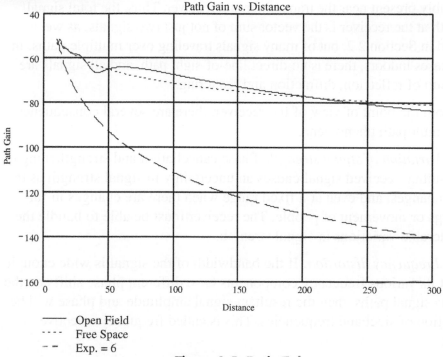

Path Gain vs. Distance

— Open Field
---- Free Space
– – Exp. = 6

Figure 2-5: Path Gain

open field distance of 300 meters, the path gain is –83 dB. The distance on the curve with exponent $n = 6$ that gives the same path gain is 34 meters. Thus, a wireless system that has an outdoor range of 300 meters may be effective only over a range of 34 meters, on the average, in an indoor installation.

The use of an empirically derived relative path loss exponent gives an estimate for average range, but fluctuations around this value should be expected. The next section shows the spread of values around the mean that occurs because of multipath radiation.

2.6 Multipath Phenomena

We have seen that reflection of a signal from the ground has a significant effect on the strength of the received signal. The nature of short- range radio links, which are very often installed indoors and use omnidirectional antennas, makes them accessible to a multitude of reflected rays, from floors, ceilings, walls, and the various furnishings and people that are invariably present near the transmitter and receiver. Thus, the total signal strength at the receiver is the vector sum of not just two signals, as we studied in Section 2.2, but of many signals traveling over multiple paths. In most cases indoors, there is no direct line-of-sight path, and all signals are the result of reflection, diffraction and scattering.

From the point of view of the receiver, there are several consequences of the multipath phenomena.

a) *Variation of signal strength*. Phase cancellation and strengthening of the resultant received signal causes an uncertainty in signal strength as the range changes, and even at a fixed range when there are changes in furnishings or movement of people. The receiver must be able to handle the considerable variations in signal strength.

b) *Frequency distortion*. If the bandwidth of the signal is wide enough so that its various frequency components have different phase shifts on the various signal paths, then the resultant signal amplitude and phase will be a function of sideband frequencies. This is called frequency selective fading.

c) *Time delay spread*. The differences in the path lengths of the various reflected signals causes a time delay spread between the shortest path and the longest path. The resulting distortion can be significant if the delay spread time is of the order of magnitude of the minimum pulse width contained in the transmitted digital signal. There is a close connection between frequency selective fading and time-delay distortion, since the shorter the pulses, the wider the signal bandwidth. Measurements in factories and other buildings have shown multipath delays ranging from 40 to 800 ns (Gibson, p. 366).

d) *Fading*. When the transmitter or receiver is in motion, or when the physical environment is changing (tree leaves fluttering in the wind, people moving around), there will be slow or rapid fading, which can contain amplitude and frequency distortion, and time delay fluctuations. The receiver AGC and demodulation circuits must deal properly with these effects.

2.7 Flat Fading

In many of the short-range radio applications covered in this book, the signal bandwidth is narrow and frequency distortion is negligible. The multipath effect in this case is classified as *flat fading*. In describing the variation of the resultant signal amplitude in a multipath environment, we distinguish two cases: (1) there is no line-of-sight path and the signal is the resultant of a large number of randomly distributed reflections; (2) the random reflections are superimposed on a signal over a dominant constant path, usually the line of sight.

Short-range radio systems that are installed indoors or outdoors in built-up areas are subject to multipath fading essentially of the first case. Our aim in this section is to determine the signal strength margin that is needed to ensure that reliable communication can take place at a given probability. While in many situations there will be a dominant signal path in addition to the multipath fading, restricting ourselves to an analysis of the case where all paths are the result of random reflections gives us an upper bound on the required margin.

Rayleigh fading

The first case can be described by a received signal $R(t)$, expressed as

$$R(t) = r \cdot \cos\left(2\pi \cdot f_c \cdot t + \theta\right) \tag{2-7}$$

where r and θ are random variables for the peak signal, or envelope, and phase. Their values may vary with time, when various reflecting objects are moving (people in a room, for example), or with changes in position of the transmitter or receiver which are small in respect to the distance between them. We are not dealing here with the large-scale path gain that is expressed in Eq. (2-5) and (2-6). For simplicity, Eq. (2-7) shows a CW (continuous wave) signal as the modulation terms are not needed to describe the fading statistics.

The envelope of the received signal, r, can be statistically described by the Rayleigh distribution whose probability density function is

$$p(r) = \frac{r}{\sigma^2} e^{\frac{-r^2}{2\sigma^2}} \tag{2-8}$$

where σ^2 represents the variance of $R(t)$ in Eq. (2-7), which is the average received signal power. This function is plotted in Figure 2-6. We normalized the curve with σ equal to 1. In this plot, the average value of the signal envelope, shown by a dotted vertical line, is 1.253. Note that it is not the most probable value, which is 1 (σ). The area of the curve between any two values of signal strength r represents the probability that the signal strength will be in that range. The average for the Rayleigh distribution, which is not symmetric, does not divide the curve area in half. The parameter that does this is the *median*, which in this case equals 1.1774. There is a 50% probability that a signal will be below the median and 50% that it will be above.

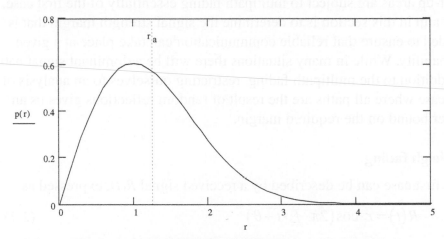

Figure 2-6: Rayleigh Probability Density Function

As stated above, the Rayleigh distribution is used to determine the signal margin required to give a desired communication reliability over a fading channel with no line of sight. The curve labeled "1 Channel" in Figure 2-7 is a cumulative distribution function with logarithmic axes. For any point on the curve, the probability of fading below the margin indicated on the abscissa is given as the ordinate. The curve is scaled such that "0 dB" signal margin represents the point where the received signal equals the mean power of the fading signal, σ^2, making the assumption that the received signal power with no fading equals the average power with fading. Some similar curves in the literature use the median power, or the power corresponding to the average envelope signal level, r_a, as the reference, "0 dB" value.

An example of using the curve is as follows. Say you require a communication reliability of 99%. Then the minimum usable signal level is that for which there is a 1% probability of fading below that level. On the curve, the margin corresponding to 1% is 20 dB. Thus, you need a signal strength 20 dB larger than the required signal if there was no fading. Assume you calculated path loss and found that you need to transmit 4 mW to allow reception at the receiver's sensitivity level. Then, to ensure that the signal will be received 99% of the time during fading, you'll need 20 dB more power or 6 dBm (4 mW) plus 20 dB equals 26 dBm or 400 mW. If you don't increase the power, you can expect loss of communication 63% of the time, corresponding to the "0 dB" margin point on the "Channel 1" curve of Figure 2-7.

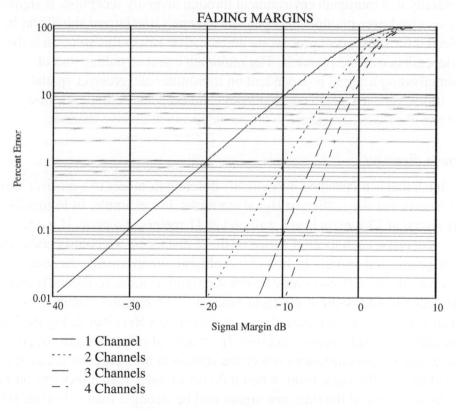

```
            1 Channel
            2 Channels
            3 Channels
            4 Channels
```

Figure 2-7: Fading Margins

The following table shows signal margins for different reliabilities.

Reliability, Percent	Fading Margin, dB
90	10
99	20
99.9	30
99.99	40

2.8 Diversity Techniques

Communication reliability for a given signal power can be increased substantially in a multipath environment through diversity reception. If signals are received over multiple, independent channels, the largest signal can be selected for subsequent processing and use. The key to this solution is the independence of the channels. The multipath effect of nulling and of strengthening a signal is dependent on transmitter and receiver spatial positions, on wavelength (or frequency) and on polarity. Let's see how we can use these parameters to create independent diverse channels.

Space diversity

A signal that is transmitted over slightly different distances to a receiver may be received at very different signal strengths. For example, in Figure 2-2 the signal at 17 meters is at a null and at 11 meters at a peak. If we had two receivers, each located at one of those distances, we could choose the strongest signal and use it. In a true multipath environment, the source, receiver, or the reflectors may be constantly in motion, so the nulls and the peaks would occur at different times on each channel. Sometimes Receiver 1 has the strongest signal, at other times Receiver 2. Figure 2-8 illustrates the paths to two receivers from several reflectors. Although there may be circumstances where the signals at both receiver locations are at around the same level, when it doesn't matter which receiver output is chosen, most of the time one signal will be stronger than the other. By selecting the strongest output, the average output after selection will be greater than the average output of one channel alone. To increase even more the probability of getting a higher average output, we could use three or more receivers. From Figure 2-7 you can find the required fading margin using diversity reception having 2, 3, or 4 channels. Note that the plots in Figure 2-7 are based on completely independent channels. When the channels are not completely independent, the results will not be as good as indicated by the plots.

It isn't necessary to use complete receivers at each location, but separate antennas and front ends must be used, at least up to the point where the signal level can be discerned and used to decide on the switch position.

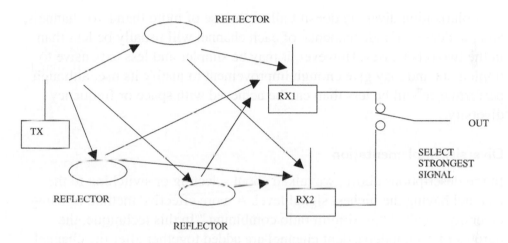

Figure 2-8: Space Diversity

Frequency diversity

You can get a similar differential in signal strength over two or more signal channels by transmitting on separate frequencies. For the same location of transmitting and receiving antennas, the occurrences of peaks and nulls will differ on the different frequency channels. As in the case of space diversity, choosing the strongest channel will give a higher average signal-to-noise ratio than on either one of the channels. The required frequency difference to get near independent fading on the different channels depends on the diversity of path lengths or signal delays. The larger the difference in path lengths, the smaller the required frequency difference of the channels.

Polarization diversity

Fading characteristics are dependent on polarization. A signal can be transmitted and received separately on horizontal and vertical antennas to create two diversity channels. Reflections can cause changes in the direction of polarization of a radio wave, so this characteristic of a signal can be used to create two separate signal channels. Thus, cross-polarized antennas can be used at the receiver only. Polarization diversity can be particularly advantageous in a portable handheld transmitter, since the orientation of its antenna will not be rigidly defined.

Polarization diversity doesn't allow the use of more than two channels, and the degree of independence of each channel will usually be less than in the two other cases. However, it may be simpler and less expensive to implement and may give enough improvement to justify its use, although performance will be less than can be achieved with space or frequency diversity.

Diversity implementation

In the descriptions above, we talked about selecting or switching to the channel having the highest signal level. A more effective method of using diversity is called "maximum ratio combining." In this technique, the outputs of each independent channel are added together after the channel phases are made equal and channel gains are adjusted for equal signal levels. Maximum ratio combining is known to be optimum as it gives the best statistical reduction of fading of any linear diversity combiner. In applications where accurate amplitude estimation is difficult, the channel phases only may be equalized and the outputs added without weighting the gains. Performance in this case is almost as good as in maximum ratio combining. [Gibson, p. 170, 171]

Space diversity has the disadvantage of requiring significant antenna separation, at least in the VHF and lower UHF bands. In the case where multipath signals arrive from all directions, antenna spacing on the order of .5λ to .8λ is adequate in order to have reasonably independent, or decorrelated, channels. This is at least one-half meter at 300 MHz. When the multipath angle spread is small—for example, when directional antennas are used—much larger separations are required.

Frequency diversity eliminates the need for separate antennas, but the simultaneous use of multiple frequency channels entails increased total power and spectrum utilization. Sometimes data are repeated on different frequencies so that simultaneous transmission doesn't have to be used. Frequency separation must be adequate to create decorrelated channels. The bandwidths allocated for unlicensed short-range use are rarely adequate, particularly in the VHF and UHF ranges (transmitting simultaneously on two separate bands can and has been done). Frequency diversity to reduce the effects of time delay spread is achieved with frequency hopping or direct sequence spread spectrum modulation, but for the spreads encountered in indoor applications, the pulse rate must be relatively high—of the order of several megabits per second—in order to be effective. For long pulse widths, the delay spread will not be a problem anyway, but multipath fading will still occur and the amount of frequency spread normally used in these cases is not likely to solve it.

When polarity diversity is used, the orthogonally oriented antennas can be close together, giving an advantage over space diversity when housing dimensions relative to wavelength are small. Performance may not be quite as good, but may very well be adequate, particularly when used in a system having portable hand-held transmitters, which have essentially random polarization.

Although we have stressed that at least two independent (decorrelated) channels are needed for diversity reception, sometimes shortcuts are taken. In some low-cost security systems, for example, two receiver antennas—space diverse or polarization diverse—are commutated directly, usually by diode switches, before the front end or mixer circuits. Thus, a minimum of circuit duplication is required. In such applications the message frame is repeated many times, so if there happens to be a multipath null when the front end is switched to one antenna and the message frames are lost, at least one or more complete frames will be correctly received when the switch is on the other antenna, which is less affected by the null. This technique works for slow fading, where the fade doesn't change much over the duration of a transmission of message frames. It doesn't appear to give any advantage during fast fading, when used with moving hand-held transmitters, for example. In that case, a receiver with one antenna will have a better chance of decoding at least one of many frames than when switched antennas are used and only half the total number of frame repetitions are available for each. In a worst-case situation with fast fading, each antenna in turn could experience a signal null.

Statistical performance measure

We can estimate the performance advantage due to diversity reception with the help of Figure 2-7. Curves labeled "2 Channels" through "4 Channels" are based on the selection combining technique.

Let's assume, as before, that we require communication reliability of 99 percent, or an error rate of 1 percent. From probability theory (see Chapter 10) the probability that two independent channels would both have communication errors is the product of the error probabilities of each channel. Thus, if each of two channels has an error probability of 10 percent, the probability that both channels will have signals below the sensitivity threshold level when selection is made is .1 times .1, which equals .01, or 1 percent. This result is reflected in the curve "2 Channels". We see that the signal margin needed for 99 percent reliability (1 percent error) is 10 dB. Using diversity reception with selection from two channels allows a reliability margin of only 10 dB instead of 20 dB, which is required if there is no diversity. Continuing the previous example, we need to transmit only 40 mW for 99 percent reliability instead of 400 mW. Required margins by selection among three channels and four channels is even less—6 dB and 4 dB, respectively.

Remember that the reliability margins using selection combining diversity as shown in Figure 2-7 are ideal cases, based on the Rayleigh fading probability distribution and independently fading channels. However, even if these premises are not realized in practice, the curves still give us approximations of the improvement that diversity reception can bring.

2.9 Noise

The ultimate limitation in radio communication is not the path loss or the fading. Weak signals can be amplified to practically any extent, but it is the noise that bounds the range we can get or the communication reliability that we can expect from our radio system. There are two sources of receiver noise—interfering radiation that the antenna captures along with the desired signal, and the electrical noise that originates in the receiver circuits. In either case, the best signal-to-noise ratio will be obtained by limiting the bandwidth to what is necessary to pass the information contained in the signal. A further improvement can be had by reducing the receiver noise figure, which decreases the internal receiver noise, but this measure is effective only as far as the noise received through the antenna is no more than about the same level as the receiver noise. Finally, if the noise can be reduced no further, performance of digital receivers can be improved by using error correction coding up to a point, which is designated as channel capacity. The capacity is the maximum information rate that the specific channel can support, and above this rate communication is impossible. The channel capacity is limited by the noise density (noise power per hertz) and the bandwidth.

Figure 2-9 shows various sources of noise over a frequency range of 20 kHz up to 10 GHz. The strength of the noise is shown in microvolts/meter for a bandwidth of 10 kHz, as received by a half-wave dipole antenna at each frequency. The curve labeled "equivalent receiver noise" translates the noise generated in the receiver circuits into an equivalent field strength so that it can be compared to the external noise sources. Receiver noise figures on which the curve is based vary in a log-linear manner with 2 dB at 50 MHz and 9 dB at 1 GHz. The data in Figure 2-9 are only a representative example of radiation and receiver noise, taken at a particular time and place.

Note that all of the noise sources shown in Figure 2-9 are dependent on frequency. The relative importance of the various noise sources to receiver sensitivity depends on their strength relative to the receiver noise. Atmospheric noise is dominant on the low radio frequencies but is not significant on the bands used for short-range communication—above around 40 MHz. Cosmic noise comes principally from the sun and from the center of our galaxy. In the figure, it is masked out by man-made noise, but in locations where man-made noise is a less significant factor, cosmic noise affects sensitivity up to 1 GHz.

Man-made noise is dominant in the range of frequencies widely used for short-range radio systems—VHF and low to middle UHF bands. It is caused by a wide range of ubiquitous electrical and electronic equipment,

Noise and its sources

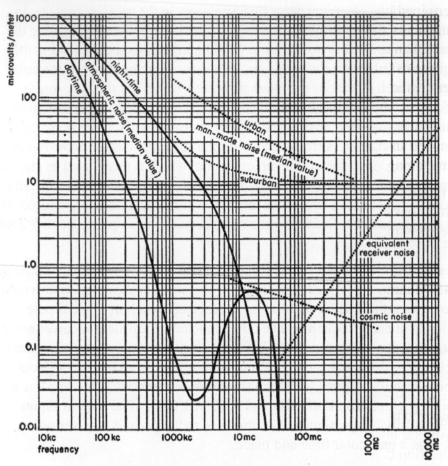

Figure 2-9: External Noise Sources
(Reference Data for Radio Engineers, Fourth Edition)

including automobile ignition systems, electrical machinery, computing devices and monitors. While we tend to place much importance on the receiver sensitivity data presented in equipment specifications, high ambient noise levels can make the sensitivity irrelevant in comparing different devices. For example, a receiver may have a laboratory measured sensitivity of –105 dBm for a signal-to-noise ratio of 10 dB. However, when measured with its antenna in a known electric field and accounting for the antenna gain, –95 dBm may be required to give the same signal-to-noise ratio.

From around 800 MHz and above, receiver sensitivity is essentially determined by the noise figure. Improved low-noise amplifier discrete components and integrated circuit blocks produce much better noise figures than those shown in Figure 2-9 for high UHF and microwave frequencies. Improving the noise figure must not be at the expense of other characteristics—intermodulation distortion, for example, which can be degraded by using a very high-gain amplifier in front of a mixer to improve the noise figure. Intermodulation distortion causes the production of inband interfering signals from strong signals on frequencies outside of the receiving bandwidth. There's more on this subject in Chapter 7.

External noise will be reduced when a directional antenna is used. Regulations on unlicensed transmitters limit the peak radiated power. When possible, it is better to use a high-gain antenna and lower transmitter power to achieve the same peak radiated power as with a lower gain antenna. The result is higher sensitivity through reduction of external noise. Manmade noise is usually less with a horizontal antenna than with a vertical antenna.

2.10 Summary

In this chapter we have looked at various factors that affect the range of reliable radio communication. Propagation of electromagnetic waves is influenced by physical objects in and near the path of line-of-sight between transmitter and receiver. We can get a first rough approximation of communication distance by considering only the reflection of the transmitted signal from the earth. If the communication system site can be classified, an empirically determined exponent can be used to estimate the path loss, and thus the range. When the transmitter or receiver is in motion, or surrounding objects are not static, the path loss varies and must be estimated statistically. We described several techniques of diversity reception that can reduce the required power for a given reliability when the communication link is subject to fading.

Noise was presented as the ultimate limitation on communication range. We saw that the noise sources to be contended with depend on the operating frequency. The importance of low-noise receiver design depends on the relative intensity of noise received by the antenna to the noise generated in the receiver.

By having some degree of understanding of electromagnetic wave propagation and noise, all those involved in the deployment of a wireless communication system—designers, installers and users—will know what performance they can expect from the system and what concrete measures can be taken to improve it.

Appendix 2-A
Maxwell's Equations

Maxwell's equations describe the relationships among five vectors that make up an electromagnetic field. These vectors are

E = electric intensity, volts/m

B = magnetic induction, Webers/m

D = Electric displacement, coulombs/m^2

H = magnetic intensity, amp/m

J = current density, amp/m^2

Since these quantities are vectors in three-dimensional space, the expressions of their relationships as differential equations or integral equations require the use of three-dimensional operators. Therefore, in order to understand and use Maxwell's equations, you must have knowledge of the subject of mathematics called vector analysis. The above quantities are also a function of time, and when they are not static, their interrelationship also involves differentiation in respect to time.

The propagation of electromagnetic waves is a result of the interdependence between electric fields and magnetic fields. Electric current flow in a wire causes a magnetic field around the wire. This is the basis of the operation of an electric motor. On the other hand, a moving wire in a magnetic field, or a changing magnetic field in the vicinity of a stationary wire, creates a potential difference in the wire and a flow of current if there is a connection between the wire ends. True, radio waves propagate in space where there are no wires. However, "displacement" currents can exist in a nonconductor, just as alternating current flows through the dielectric material of a capacitor.

We state here the four basic laws of electromagnetism in their simplified integral form. They each also have an equivalent form using three-dimensional derivative operators. In the equations, overhead lines indicate three-dimensional vectors. To make some equations a little easier to understand, we use familiar scalar quantities instead of vectors listed above. Electrical current is referred to in the first equation as part of total current instead of the surface integral of the current density $\boldsymbol{J}$. Similarly the magnetic flux ϕ is referred to in equation (2) instead of the integral of flux density $\boldsymbol{B}$.

1)

$$\oint \overline{H} \bullet d\overline{s} = I_{total}$$

This law describes the connection between magnetic intensity and electric current. On the left is a closed line integral of magnetic intensity around a path that encloses a flow of current. This current, I_{total}, has two components: a conduction current such as current through a wire, and a displacement current that is produced by an electric charge that is changing with time. In the case of propagation through space, I_{total} is only displacement current since conductors are absent.

2)

$$\oint \overline{E} \bullet d\overline{s} = -\frac{\partial \phi}{\partial t}$$

This equation complements the first equation above. It shows that the electrical intensity integrated on a closed path equals the negative value of the rate of change of magnetic flux flowing through the area enclosed by that path. (A partial derivative symbol is used on the right since ϕ is also a function of its spatial coordinates, which are constant here.) Compare this equation with Faraday's law of magnetic induction

$$V_{out} = -N\frac{d\phi}{dt}$$

which is the basis for operation of an electric generator (N is the number of turns of a coil exposed to changing magnetic flux ϕ).

3)

$$\oiint \overline{D} \bullet d\overline{S} = Q$$

A net electric charge Q gives rise to an electric flux density D over a surface totally enclosing Q. The value of the integral of D over this surface equals the value of Q. This equation connects an electric charge with the electric field surrounding it. In space there is no net charge so the electric flux density integrated around a closed surface (such as the surface of a sphere) must be zero.

4)

$$\oiint \overline{B} \bullet d\overline{S} = 0$$

Here we have the complement of the previous equation. An isolated magnetic pole or magnetic charge doesn't exist, in contrast to the isolated electric charge of equation 3). Magnetic flux lines are continuous loops, so lines that enter a closed surface must also come out of it. The result is that the integral of magnetic flux density B over any closed surface, expressed on the left side of this equation, is zero. There's no such thing as a magnet having only a north pole from which may emanate magnetic flux, never to return.

For an excellent introduction to Maxwell's equations, see the tutorial by George J. Spix, Reference 54.

CHAPTER 3

Antennas and Transmission Lines

3.1 Introduction

The antenna is the interface between the transmitter or the receiver and the propagation medium, and it therefore is a deciding factor in the performance of a radio communication system. The principal properties of antennas—directivity, gain, and radiation resistance—are the same whether referred to as transmitters or receivers. The principle of reciprocity states that the power transferred between two antennas is the same, regardless of which is used for transmission or reception, if the generator and load impedances are conjugates of the transmitting and receiving antenna impedances in each case.

First we define the various terms used to characterize antennas. Then we discuss several types of antennas that are commonly used in short-range radio systems. Finally, we review methods of matching the impedances of the antenna to the transmitter or receiver RF circuits.

3.2 Antenna Characteristics

Understanding the various characteristics of antennas is a first and most important step before deciding what type of antenna is most appropriate for a particular application. While antennas have several electrical characteristics, often a primary concern in choosing an antenna type is its physical size. Before dealing with the various antenna types and the shapes and sizes they come in, we first must know what the antenna has to do.

Antenna impedance

As stated in the introduction, the antenna is an interface between circuits and space. It facilitates the transfer of power between space and the transmitter or receiver. The antenna impedance is the load for the transmitter or the input impedance to the receiver. It is composed of two parts—radiation resistance and ohmic resistance. The radiation resistance is a virtual resistance that, when multiplied by the square of the RMS current in the antenna at its feed point, equals the power radiated by the antenna in the case of a transmitter or extracted from space in the case of a receiver. It is customary to refer the radiation resistance to a current maximum in the case of an ungrounded antenna, and to the current at the base of the antenna when the antenna is grounded. Transmitter power delivered to an antenna will always be greater than the power radiated. The difference between the transmitter power and the radiated power is power dissipated in the ohmic resistance of the antenna conductor and in other losses. The efficiency of an antenna is the ratio of the radiated power to the total power absorbed by the antenna. It can be expressed in terms of the radiation resistance R_r and loss resistance R_l as

$$\text{Antenna Efficiency } (\%) = 100 \times (R_r / (R_r + R_l)) \qquad (3\text{-}1)$$

The resistance seen by the transmitter or receiver at the antenna terminals will be equal to the radiation resistance plus the loss resistance only if these terminals are located at the point of maximum current flow in the antenna. The impedance at this point may have a reactive component, too. When there is no reactive component, the antenna is said to be resonant. Maximum power transfer between the antenna and transmitter or receiver will occur only when the impedance seen from the antenna terminals is the complex conjugate of the antenna impedance.

It is important to match the transmitter to the antenna not only to get maximum power transfer. Attenuation of harmonics relative to the fundamental frequency is maximized when the transmitter is matched to the antenna—an important point in meeting the spurious radiation requirements for license-free transmitters. The radiation resistance depends on the proximity of the antenna to conducting and insulating objects. In particular, it depends on the height of the antenna from the ground. Thus, the antenna-matching circuit of a transmitter with integral antenna that is intended to be hand-held should be optimized for the antenna impedance in a typical operating situation.

Directivity and gain

The directivity of an antenna relates to its radiation pattern. An antenna which radiates uniformly in all directions in three-dimensional space is called an isotropic antenna. Such an antenna doesn't exist, but it is convenient to refer to it when discussing the directional properties of an antenna. All real antennas radiate stronger in some directions than in others. The directivity of an antenna is defined as the power density of the antenna in its direction of maximum radiation in three-dimensional space divided by its average power density. The directivity of the hypothetical isotropic radiator is 1, or 0 dB. The directivity of a half-wave dipole antenna is 1.64, or 2.15 dB.

The radiation pattern of a wire antenna of short length compared to a half wavelength is shown in Figure 3-1a. The antenna is high enough so as not to be affected by the ground. If the antenna wire direction is parallel to the earth, then the pattern represents the intersection of a horizontal plane with the solid pattern of the antenna shown in Figure 3-1b. A vertical wire antenna is omnidirectional; that is, it has a circular horizontal radiation pattern and directivity in the vertical plane.

The gain of an antenna is the directivity times the antenna efficiency. When antenna losses are low, the two terms are almost the same. In general, when you are interested in the directional discrimination of an antenna, you will be interested in its directivity. Gain is used to find the maximum radiated power when the power into the antenna is known.

Effective area

Another term often encountered is the effective area of an antenna. Wave propagation can be described as if all of the radiated power is spread over the surface of a sphere whose area expands according to the square of the distance (in free space). The power captured by the receiving antenna is then the capture area, or effective area, of the antenna times the power density at that location. The power density is the radiated power divided by the surface area of the sphere.

Figure 3-1: Short Dipole Antenna (ARRL Antenna Book)

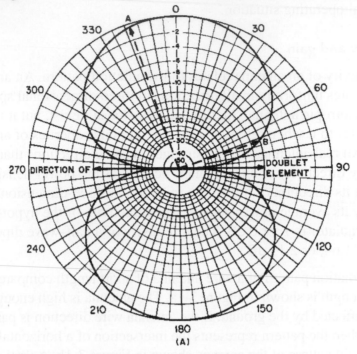

a) Directive pattern in plane containing antenna

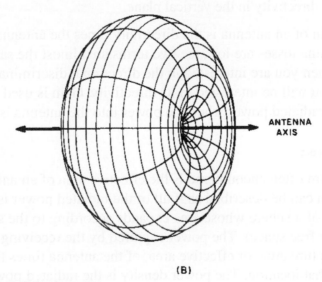

b) Solid Diagram

Courtesy *Antenna Book, 16th Edition*, published by ARRL.

The effective area of an antenna related to gain and wavelength is shown in the following expression:

$$A_e = \frac{\lambda^2 \cdot G}{4\pi}$$ (3-2)

This expression shows us that the capture ability of an antenna of given gain G grows proportionally as the square of the wavelength. The antenna of a particular configuration captures less power at higher frequencies. (Remember that frequency is inversely proportional to wavelength λ.)

When the electric field strength E is known, the power density is

$$P_d = \frac{E^2}{120\pi}$$ (3-3)

Thus, received power can be found when field strength is known by multiplying (3-2) times (3-3):

$$P_R = \frac{E^{2}\lambda^2 G}{480\pi^2}$$ (3-4)

It's intuitive to note that the effective antenna area has some connection with the physical size of the antenna. This is most obvious in the case of microwave antennas where the effective area approaches the physical aperture. From (3-4) it appears that for a given radiated power and thus field strength, the lower frequency (longer wavelength) systems will give stronger receiver signals than high-frequency equipment. However, short-range devices are often portable or are otherwise limited in size and their antennas may have roughly the same dimensions, regardless of frequency. The lower frequency antennas whose sizes are small fractions of a wavelength have poor efficiency and low gain and therefore may have effective areas similar to their high-frequency counterparts. Thus, using a low frequency doesn't necessarily mean higher power at the receiver, which Eq. (3-4) may lead us to believe.

Polarization

Electromagnetic radiation is composed of a magnetic field and an electric field. These fields are at right angles to each other, and both are in a plane normal to the direction of propagation. The direction of polarization refers to the direction of the electric field in relation to the earth. Linear polarization is created by a straight wire antenna. A wire antenna parallel to the earth is horizontally polarized and a wire antenna normal to the earth is vertically polarized.

The electric and magnetic fields may rotate in their plane around the direction of propagation, and this is called elliptical polarization. It may be created by perpendicular antenna elements being fed by coherent RF signals that are not in the same time phase with each other. Circular polarization results when these elements are fed by equal power RF signals which differ in phase by 90°, which causes the electric (and magnetic) field to make a complete 360° rotation every period of the wave (a time of 1/freq. seconds). Some antenna types, among them the helical antenna, produce elliptic or circular polarization inherently, without having two feed points. There are two types of elliptical polarization, right hand and left hand, which are distinguished by the direction of rotation of the electric field.

The polarization of a wave, or an antenna, is important for several reasons. A horizontally polarized receiving antenna cannot receive vertically polarized radiation from a vertical transmitting antenna, and vice versa. Similarly, right-hand and left-hand circular antenna systems are not compatible. Sometimes this quality is used to good advantage. For example, the capacity of a microwave link can be doubled by transmitting two different information channels between two points on the same frequency using oppositely polarized antenna systems.

The degree of reflection of radio signals from the ground is affected by polarization. The phase and amount of reflection of vertically polarized waves from the ground are much more dependent on the angle of incidence than horizontally polarized waves.

Except for directional, line-of-sight microwave systems, the polarity of a signal may change during propagation between transmitter and receiver. Thus, in most short-range radio applications, a horizontal antenna will receive transmissions from a vertical antenna, for example, albeit with some attenuation. The term *cross polarization* defines the degree to which a transmission from an antenna of one polarization can be received by an antenna of the opposite polarization. Often, the polarization of a transmitter or receiver antenna is not well defined, such as in the case of a handheld device. A circular polarized antenna can be used when the opposite antenna polarization is not defined, since it does not distinguish between the orientation of the linear antenna.

Bandwidth

Antenna bandwidth is the range of frequencies over which the antenna can operate while some other characteristic remains within a given range. Very frequently, the bandwidth is related to the antenna impedance expressed as standing wave ratio. Obviously, a device that must operate over a number of frequency channels in a band must have a comparatively wide bandwidth antenna. Less obvious are the bandwidth demands for a single frequency device.

A narrow bandwidth or high Q antenna will discriminate against harmonics and other spurious radiation and thereby will reduce the requirements for a supplementary filter, which may be necessary to allow meeting the radio approval specifications. On the other hand, drifting of antenna physical dimensions or matching components could cause the power output (or sensitivity) to fall with time. Changing proximity of nearby objects or the "hand effect" of portable transmitters can also cause a reduction of power or even a pulling of frequency, particularly in low-power transmitters with a single oscillator stage and no buffer or amplifier stage.

Antenna factor

The antenna factor is commonly used with calibrated test antennas to make field strength measurements on a test receiver or spectrum analyzer. It relates the field strength to the voltage across the antenna terminals when the antenna is terminated in its specified impedance (usually 50 or 75 ohms):

$$AF = \frac{E}{V}$$

(3-5)

where $AF =$ antenna factor in (meters)$^{-1}$

$E =$ field strength in V/m

$V =$ load voltage in V

Usually the antenna factor is stated in dB:

$$AF_{dB}(m^{-1}) = 20 \log (E/V)$$

The relationship between numerical gain G and antenna factor AF is:

$$AF = \frac{4\pi}{\lambda} \cdot \sqrt{\frac{30}{R_L \cdot G}} \tag{3-6}$$

where R_L is the load resistance, usually 50 ohms.

3.3 Types of Antennas

In this section we review the characteristics of several types of antennas that are used in short-range radio devices. The size of the antenna is related to the wavelength, which in turn is found when frequency is known from

wavelength = (velocity of propagation)/(frequency)

The maximum velocity of propagation occurs in a vacuum. It is approximately 300,000,000 meters/second, with little difference in air. This figure is less in solid materials, so the wavelength will be shorter for antennas printed on circuit board materials or protected with a plastic coating.

Dipole

The dipole is a wire antenna fed at its center. The term usually refers to an antenna whose overall length is one-half wavelength. In free space its radiation resistance is 73 ohms, but that value will vary somewhat when the ground or other large conducting objects are within a wavelength distance from it. The dipole is usually mounted horizontally, but if mounted vertically, its transmission line feeder cable should extend from it at a right angle for a distance of at least a quarter wavelength.

In free space, the radiation pattern of a horizontal half-wave dipole at zero degrees elevation is very similar to that of the dipole shown in Figure 3-1a. It has a directivity of 1.64, or 2.15 dB. It would seem that this antenna would not be usable for short-range devices that need an omnidirectional radiation, but this isn't necessarily the case. The radiation pattern at an elevation of 30 degrees when the antenna is a half wavelength above a conductive plate is shown in Figure 3-2. The radiation is 8 dB down from maximum along the direction of the wire. When used indoors where there are a multitude of reflections from walls, floor and ceiling, the horizontal dipole can give good results in all directions.

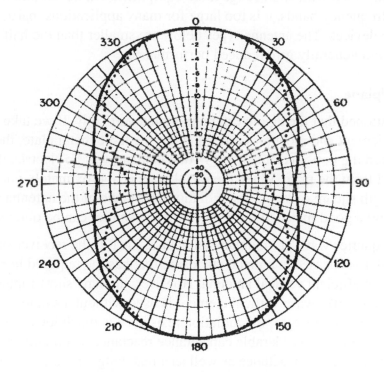

Figure 3-2: Dipole Pattern at 30° Elevation (solid line)

Courtesy 1990 *ARRL Handbook.*

The half-wave dipole antenna is convenient to use because it is easy to match a transmitter or receiver to its radiation resistance. It has high efficiency, since wire ohmic losses are only a small fraction of the radiation resistance. Also, the antenna characteristics are not much affected by the size or shape of the device it is used with, and it doesn't use a ground plane. Devices whose dimensions are small relative to the antenna size can directly feed the dipole, with little or no transmission line. For increased compactness, the two antenna elements can be extended at an angle instead of being in a straight line.

In spite of its many attractive features, the half-wave dipole is not commonly used with short-range radio equipment. On the common unlicensed frequency bands, it is too large for many applications, particularly portable devices. The antenna types below are smaller than the half-wave dipole, and generally give reduced performance.

Groundplane

We mentioned that the dipole can be mounted vertically. If we take one dipole element and mount it perpendicular to a large metal plate, then we don't need the bottom element—a virtual element will be electrically reflected from the plate. When the metal plate is approximately one-half wavelength square or larger, the radiation resistance of the antenna is 36 ohms, and a good match to the receiver or transmitter can be obtained.

The quarter-wave groundplane antenna is ideal if the receiver or transmitter is encased in a metal enclosure that has the required horizontal area for an efficient vertical antenna. However, in many short-range devices, a quarter-wave vertical element is used without a suitable groundplane. In this case, the radiation resistance is much lower than 36 ohms and there is considerable capacitance reactance. An inductor is needed to cancel the reactance as well as a matching circuit to assure maximum power transfer between the antenna and the device. The ohmic losses in the inductor and other matching components, together with the low radiation resistance, result in low antenna efficiency. When possible, the antenna length should be increased to a point where the antenna is resonant, that is, has no reactance. The electrical length can be increased and capacitive reactance reduced by winding the bottom part of the antenna element into a coil having several turns. In this way, the loss resistance may be reduced and efficiency increased.

Loop

The loop antenna is popular for hand-held transmitters particularly because it can be printed on a small circuit board and is less affected by nearby conducting objects than other small resonant antennas. Its biggest drawback is that it is very inefficient.

A loop antenna whose dimensions are small compared to a wavelength—less than 0.1λ—has essentially constant current throughout. Its radiation field is expressed as

$$E(\theta) = \frac{120\pi^2 \cdot I \cdot N \cdot A}{r \cdot \lambda^2} \cdot \cos\theta \qquad (3\text{-}7)$$

where

I is its current

A is the loop area

r is the distance

θ is the angle from the plane of the loop

N is the number of turns

From this expression we can derive the radiation resistance which is

$$R_r = 320 \cdot \pi^4 \cdot \frac{(A \cdot N)^2}{\lambda^4} \qquad (3\text{-}8)$$

Loop antennas are frequently used in small hand-held remote control transmitters on the low UHF frequencies. The radiation resistance is generally below a tenth of an ohm and the efficiency under 10%. In order to match the transmitter output stage to the low antenna resistance, parallel resonance is created using a capacitor across the loop terminals. While it may appear that the radiation resistance and hence the efficiency could be raised by increasing the number of turns or the area of the loop, the possibilities with this approach are very limited. Increasing the area increases the loop inductance, which requires a smaller value of resonating capacitance. The limit on the area is reached when this capacitance is several picofarads, and then we get the radiation resistance and approximate efficiency as mentioned above.

Because of the low efficiency of the loop antenna, it is rarely used in UHF short-range receivers. An exception is pager receivers, which use low data rates and high sensitivity to help compensate for the low antenna efficiency. One advantage of the loop antenna is that it doesn't require a ground plane.

In low-power unlicensed transmitters, the low efficiency of the loop is not of much concern, since it is the radiated power that is regulated, and at the low powers in question, the power can be boosted enough to make up for the low efficiency. A reasonably high Q is required in the loop circuit, however, in order to keep harmonic radiation low in respect to the fundamental. In many short-range transmitters, it is the harmonic radiation specification that limits the fundamental power output to well below the allowed level.

For detailed information on loop antenna performance and matching, see References 15 and 58.

Example

We will design a loop antenna for a transmitter operating on 315 MHz. The task is easy using the Mathcad worksheet "Loop Antenna."

Given data: f = 315 MHz; G10 circuit board 1/16" thick, 1 oz. copper plating, and dielectric constant = 4.7; loop sides 25 mm and 40 mm, conductor width 2 mm.

Enter the relevant data in the worksheet.

The results of this example are:

    Radiation resistance = .038 ohm

    Loss resistance = 0.15 ohm

    Efficiency = 20.1% = –7 dB

    Resonating capacitance 3.65 pF

The results from using the loop antenna worksheet are not particularly accurate, but they do give a starting point for design. Efficiency can be expected to be worse than that calculated because circuit board losses were not accounted for, nor were the effects of surrounding components. There will also be significant losses in the matching circuit because of the difficulty of matching the high output impedance of the low-power transmitter to the very low impedance of the loop. Transmitters designed to operate from low battery voltages can be expected to be better in this respect (see "Transmitter Output Impedance" later in this section).

Helical

The helical antenna can give much better results than the loop antenna, when radiation efficiency is important, while still maintaining a relatively small size compared to a dipole or quarter-wave ground plane.

The helical antenna is made by winding stiff wire in the form of a spring, whose diameter and pitch are very much smaller than a wavelength, or by winding wire on a cylindrical form. See Figure 3-3. This helical winding creates an apparent axial velocity along the "spring" which is much less than the velocity of propagation along a straight wire approximately the speed of light in space. Thus, a quarter wave on the helical spring will be much shorter than on a straight wire. The antenna is resonant for this length, but the radiation resistance will be lower and consequently the efficiency is less than that obtained from a standard quarter-wave antenna. The helical antenna resonates when the wire length is in the neighborhood of a half wavelength. Impedance matching to a transmitter or receiver is relatively easy.

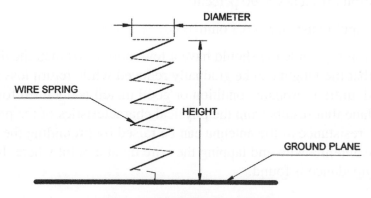

Figure 3-3: Helical antenna

The radiating surface of the helical antenna has both vertical and horizontal components, so its polarization is elliptic. However, for the form factors most commonly used, where the antenna length is several times larger than its diameter, polarization is essentially vertical.

The helical antenna should have a good ground plane for best and predictable performance. In hand-held devices, the user's arm and body serve as a counterpoise, and the antenna should be designed for this configuration.

The Mathcad worksheet "Helical Antenna" helps design a helical antenna. We'll demonstrate by an example.

Example

Our antenna will be designed for 173 MHz. We will wind it on a 10-mm form with AWG 20 wire. We want to find the number of turns to get a resonant antenna 16 cm high. We also want an approximation of the radiation resistance and the antenna efficiency.

Given: The mean diameter of the antenna D = 10.8 mm (includes the wire diameter). Wire diameter of AWG 20 is d = .8 mm. Antenna height h = 160 mm. Frequency f = 173 MHz.

We insert these values into the helical antenna worksheet and get the following results:

Number of turns = 26

Wire length = 89 cm = .514 λ

Radiation efficiency = 90 percent

Total input resistance = 6.1 ohm

The prototype antenna should have a few more turns than the design value so that the length can be gradually reduced while return loss is monitored, until a resonant condition or good match is obtained for the ground plane that results from the physical characteristics of the product. The input resistance of the antenna can be raised by grounding the bottom end of the antenna wire and tapping the wire up at a point where the desired impedance is found.

Patch

The patch antenna is convenient for microwave frequencies, specifically on the 2.4-GHz band and higher. It consists of a plated geometric form (the patch) on one side of a printed circuit board, backed up on the opposite board side by a groundplane plating which extends beyond the dimensions of the radiating patch. Rectangular and circular forms are the most common, but other shapes—for example, a trapezoid—are sometimes used. Maximum radiation is generally perpendicular to the board. A square half-wave patch antenna has a directivity of 7 to 8 dB.

A rectangular patch antenna is shown in Figure 3-4. The dimension L is approximately a half wavelength, calculated as half the free space wavelength (λ) divided by the square root of the effective dielectric constant (ε) of the board material. It must actually be slightly less than a half wavelength because of the fringing effect of the radiation from the two opposite edges that are L apart. As long as the feed is on the centerline, the two other edges don't radiate. The figure shows a microstrip feeder, which is convenient because it is etched on the board together with the patch and other component traces on the same side.

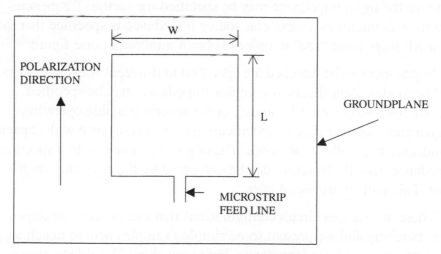

Figure 3-4: Patch Antenna

The impedance at the feed point depends on the width *W* of the patch. A microstrip transforms it to the required load (for transmitter) or source (for receiver) impedance. The feed point impedance can be made to match a transmission line directly by moving the feed point from the edge on the centerline toward the center of the board. In this way a 50-ohm coax transmission line can be connected directly to the underside of the patch antenna, with the center conductor going to the feed point through a via and the shield soldered to the groundplane.

The Mathcad "Patch Antenna" worksheet on the enclosed CDROM helps design a rectangular patch antenna. It includes calculations for finding the coax cable feed point location.

3.4 Impedance Matching

Impedance matching is important in transmitters and receivers for getting the best transfer of power between the antenna and the device. In a receiver, matching is often done in two stages—matching the receiver input to 50 ohms to suit a bandpass filter and to facilitate laboratory sensitivity measurements, and then matching from 50 ohms to the antenna impedance. Receiver modules most often have 50 ohms input impedance. Receiver integrated circuits or low-noise RF amplifiers may have 50 ohms input, or the input impedance may be specified for various frequencies of operation. Sometimes a particular source impedance is specified that the input RF stage must "see" in order to obtain minimum noise figure.

Impedances to be matched are specified in different ways in component or module data sheets. A complex impedance may be specified directly for the operating frequency or for several possible operating frequencies. Another type of specification is by a resistance with capacitor or inductor in parallel or in series. The degree of matching to a specified impedance, usually 50 ohms, can be expressed by the *reflection coefficient*. This will be discussed later.

There are various circuit configurations that can be used for impedance matching and we present some simple examples here to match a pure resistance to a complex impedance. First, you should be able to express an impedance or resistance-reactance combination in parallel or serial form, whichever is convenient for the matching topography you wish to use. You can do this using the Mathcad worksheet "Impedance Transformations." Then use the worksheet "Impedance Matching" to find component values to match a wide range of impedances. The parallel or series source reactance must be separated from the total adjacent derived reactance value to get the value of the component to use in the matching circuit. Example 1 demonstrates this for a parallel source capacitor.

Remember that coils and capacitors are never purely reactive. The losses in coils, specified by the quality factor Q, are often significant whereas those in capacitors are usually ignored. In a parallel equivalent circuit (loss resistance in parallel to the reactance), $Q = R/X$ (X is the reactance). In a series equivalent circuit, $Q = X/R$. If the loss resistance is within a factor of up to around 5 of a resistance to be matched, it should be combined with that resistance before using the impedance matching formula. Example 2 shows how to do it.

Use nearest standard component values for the calculated values. Variable components may be needed, particularly in a high Q circuit. Remember also that stray capacitance and inductance will affect the matching and should be considered in selecting the matching circuit components.

Example 1

Figure 3-5 shows a circuit that can be used for matching a high impedance, such as may be found in a low-power transmitter, to 50 ohms.

Let's use it to match a low-power transmitter output impedance of 1000 ohms (see the section "Transmitter Output Impedance") and 1.5-pF parallel capacitance to a 50-ohm bandpass filter or antenna. The frequency is 315 MHz. We use "Impedance Matching" worksheet, circuit 3.

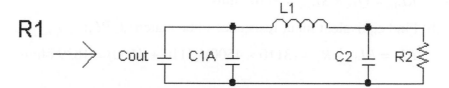

Figure 3-5: Impedance Transformation, Example 1

Given values are:

$R_1 = 1000$ ohms $C_{out} = 1.5$ pF $R_2 = 50$ ohms $f = 315$ MHz

(a) At the top of the worksheet, set $f = 315$ MHz. Under circuit (3) set R_1 and R_2 to 1000 and 50 ohms, respectively. Select a value for Q.

$Q = 10$

(b) Rounded off results are:

$C_1 = 5.1$ pF

$C_2 = 20.3$ pF

$L_1 = 101$ nH

(c) C_1 of the worksheet is made up of the parallel combination of C_{out} and C_{1A} of Figure 3-5:

$C_{1A} = C_1 - C_{out} = 5.1$ pF $- 1.5$ pF $= 3.6$ pF

Example 2

We want to match the input of an RF mixer and IF amplifier integrated circuit (such as Philips NE605) to a 50-ohm antenna at 45 MHz. The equivalent input circuit is 4500 ohms in parallel with 2.5 pF. We choose to use a parallel coil L_{1A} having a value of 220 nH and a Q_{L1A} of 50. See Figure 3-6.

Given: $f = 45$ MHz, $R_1 = 50$ ohms, $R_{in} = 4.5$K ohms, $C_{in} = 2.5$ pF, $L_{1A} = 220$ nH, $Q_{1A} = 50$

Find: C_1 and C_2

(a) Calculate RL_{1A}

$XL_{1A} = 62.2$ ohms (you can use "Conversions" worksheet)

$RL_{1A} = Q_{1A} \times XL_{1A} = 3110$ ohms

(b) Find equivalent input resistance to be matched, RL_1:

$RL_1 = RL_{1A} \parallel R_{in} = (3110 \times 4500)/(3110 + 4500) = 1839$ ohms

(c) Find equivalent parallel inductance L_1

$XC_{in} = -1415$ ohms ("Conversions" worksheet)

$XL_1 = XL_{1A} \parallel XC_{in} = (62.2 \times (-1415))/(62.2 - 1415) = 65.06$ ohms

(d) Find Q, which is needed for the calculation of C_1 and C_2 using the "Impedance Matching" worksheet:

$Q = RL_1/XL_1 = 28.27$

(e) Use the worksheet "Impedance Matching" circuit (4) to find C_1 and C_2, after specifying R_1, R_2, and Q:

$R_1 = 50$ ohms $R_2 = RL_1 = 1839$ ohms $Q = 28.27$

Results:

$C_1 = 65$ pF

$C_2 = 322$ pF

It may seem that the choice of the parallel inductor was arbitrary, but that's the designer's prerogative, as long as the resultant Q is greater than the minimum Q given in the worksheet example (in this case approximately 6). The choice of inductance determines the circuit Q, and consequently the bandwidth of the matching circuit. The total Q of the circuit includes the loading effect of the source resistance. Its value is one half the Q used in the design procedure, or $28.27/2 =$ approximately 14 in this example.

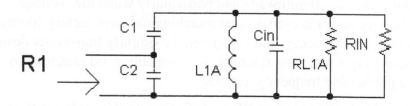

Figure 3-6: Impedance Transformation, Example 2

Example 3

We have a helical antenna with 15-ohm impedance that will be used with a receiver module having a 50-ohm input. The operating frequency is 173 MHz. The matching network is shown in Figure 3-7.

Given: f = 173 MHz, R_1 = 50 ohms, R_2 = 15 ohms

Use these values in the "Matching Impedance" worksheet, circuit (1), to get the matching network components:

$$L_1 = 21.1 \text{ nH}$$

$$C_1 = 28.1 \text{ pF}$$

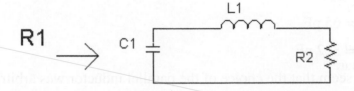

Figure 3-7: Impedance Transformation, Example 3

Transmitter output impedance

In order to get maximum power transfer from a transmitter to an antenna, the RF amplifier output impedance must be known, as well as the antenna impedance, so that a matching network can be designed as shown in the previous section. For very low-power transmitters with radiated powers of tens of microwatts at the most, close matching is not critical. However, for a radiated power of 10 milliwatts and particularly when low-voltage lithium battery power is used, proper matching can save battery energy due to increased efficiency, and can generally simplify transmitter design. Besides, an output bandpass filter needs reasonably good matching to deliver a predictable frequency response.

A simplified estimate of an RF amplifier's output impedance R_L is given by the following expression:

$$R_L = \frac{\left(V_{CC} - V_{CE(sat)}\right)^2}{2P} \tag{3-9}$$

V_{CC} is the supply voltage to the RF stage, $V_{CE(sat)}$ is the saturation voltage of the RF transistor at the operating frequency, and *P* is the power output.

Transmission lines

In many short-range radio devices the transmitter or receiver antenna is an integral part of the device circuitry and is coupled directly to the transmitter output or receiver input circuit through discrete components. This is particularly the case with portable equipment. Devices with an external antenna located away from the equipment housing need a transmission line to connect the antenna to the input or output circuit. The transmission line is an example of a *distributed circuit* and it affects the coupling or transfer of the RF signal between the device RF circuit and the antenna. At high UHF and microwave frequencies, even the short connection between an internal antenna and the RF circuit is considered a transmission line whose characteristics must be designed to achieve proper impedance matching.

The transmission line can take several forms, among them coaxial cable, balanced two-wire cable, microstrip, and waveguide (a special case, not considered below).

A basic characteristic of a transmission line is its *characteristic impedance*. Its value depends on the capacitance per unit length *C* and inductance per unit length *L*, which in turn are functions of the physical characteristics of the line and the dielectric constant of the material surrounding the conductors. In the ideal case when there are no losses in the line the relationship is

$$Z_0 = \sqrt{\frac{L}{C}}$$

where *L* is the inductance per unit length in henrys and *C* is the capacitance per unit length in farads. Another important characteristic is the velocity factor, which is the ratio of the propagation velocity, or phase velocity, of the wave in the line to the speed of light. The velocity factor depends on the dielectric constant, ε, of the material enclosing the transmission line conductors as

$$VF = \frac{1}{\sqrt{\varepsilon}}$$

Also important in specifying a transmission line, particularly at VHF and higher frequencies and relatively long lines, is the attenuation or line loss.

For example, the characteristics of a commonly used coaxial cable, RG-58C, are:

Characteristic impedance	50 ohms
Inductance per meter	0.25 microhenry
Capacitance per meter	101 picofarad
Velocity factor	0.66
Attenuation at 50 MHz	3.5 dB per 30 meters
Attenuation at 300 MHz	10 dB per 30 meters

In the previous section we talked about matching the antenna impedance to the circuit impedance. When the antenna is connected to the circuit through a transmission line, the impedance to be matched, seen at the circuit end of the transmission line, may be different from the impedance of the antenna itself. It depends on the characteristic impedance of the transmission line and the length of the line.

Several terms which define the degree of impedance matching, usually relating to transmission lines and antennas, are presented below.

Standing Wave Ratio is a term commonly used in connection with matching a transmission line to an antenna. When the load impedance differs from the characteristic impedance of the transmission line, the peak voltage on the line will differ from point to point. On a line whose length is greater than a half wavelength, the distance between voltage peaks or between voltage nulls is one-half wavelength. The ratio of the voltage peak to the voltage null is the standing wave ratio, abbreviated *SWR*, or *VSWR* (voltage standing wave ratio). The ratio of the peak current to the minimum current is the same as the *VSWR*, or *SWR*.

When the load is a pure resistance, R, and is larger than the characteristic impedance of the line (also considered to be a pure resistance), we have

$$SWR = R/Z_0$$

If the load resistance is less than the line characteristic impedance, then

$$SWR = Z_0/R$$

In the general case where both the line impedance and the characteristic impedance may have reactance components, thus being complex, the voltage standing wave ratio is

$$SWR = \frac{1 + \left| \dfrac{Z_{load} - Z_0}{Z_{load} + Z_0} \right|}{1 - \left| \dfrac{Z_{load} - Z_0}{Z_{load} + Z_0} \right|} \qquad (3\text{-}10)$$

Reflection Coefficient is the ratio of the voltage of the reflected wave from a load to the voltage of the forward wave absorbed by the load:

$$\rho = \frac{E_r}{E_f}$$

When the load is perfectly matched to the transmission line, the maximum power available from the generator is absorbed by the load, there is no reflected wave, and the reflection coefficient is zero. For any other load impedance, less power is absorbed by the load, and what remains of the available power is reflected back to the generator. When the load is an open or short circuit, or a pure reactance, all of the power is reflected back, the reflected voltage equals the forward voltage, and the reflection coefficient is unity. We can express the reflection coefficient in terms of the load impedance and characteristic impedance as

$$\rho = \frac{Z_{load} - Z_0}{Z_{load} + Z_0} \qquad (3\text{-}11)$$

The relation between the standing wave ratio and the reflection coefficient is

$$SWR = \frac{1+|\rho|}{1-|\rho|}$$ (3-12)

Return Loss is an expression of the amount of power returned to the source relative to the available power from the generator. It is expressed in decibels as

$$RL = -20\log(|\rho|)$$ (3-13)

Note that the return loss is always equal to or greater than zero.

Of the three terms relating to transmission line matching, the reflection coefficient gives the most information, since it is a complex number. As for the other two terms, *SWR* may be more accurate for large mismatches, whereas return loss presents values with greater resolution than *SWR* when the load impedance is close to the characteristic impedance of the line.

A plot of forward and reflected powers for a range of *SWRs* is given in Figure 3-8. This plot is convenient for seeing the effect of an impedance mismatch on the power actually dissipated in the load or accepted by the antenna, which is the forward power minus the reflected power.

Transmission line losses are not represented in the above definitions. Their effect is to reduce the *SWR* and increase the return loss, compared to a lossless line with the same load. This may seem to contradict the expressions given, which are in terms of load impedance, but that is not so. For instance, the load impedance in the expression for *SWR* (equation (3-10)) is *the impedance at a particular point on the line where the SWR is wanted* and not necessarily the impedance at the end of the line. Thus, a long line with high losses may have a low *SWR* measured at the generator end, but a high *SWR* at the load. Transmission line loss is specified for a perfectly matched line, but when a mismatch exists, the loss is higher because of higher peak current and a resulting increased I^2R power dissipation in the line.

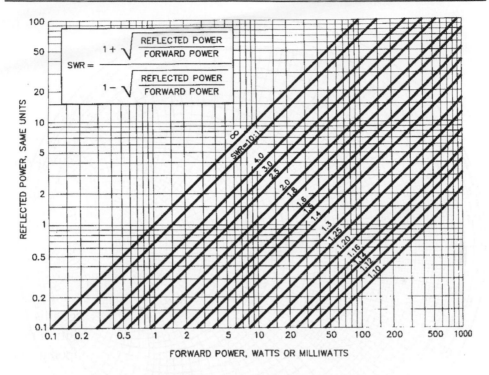

Figure 3-8: SWR, Forward and Reflected Power
Courtesy *Antenna Book, 16ᵗʰ Edition*, ARRL

Smith chart

A convenient tool for finding impedances in transmission lines and designing matching networks is the Smith chart, shown in Figure 3-9.

The Smith chart is a graph on which you can plot complex impedances and admittances (admittance is the inverse of impedance). An impedance value on the chart is the intersection of a resistance circle, labeled on the straight horizontal line in the middle, and a reactance arc, labeled along the circumference of the "0" resistance circle. Figure 3-10 gives an expanded view of the chart with some of the labels. The unique form of the chart was devised for convenient graphical manipulation of impedances and admittances when designing matching networks, particularly when transmission lines are involved. The Smith chart is useful for dealing with distributed parameters which describe the characteristics of circuit board traces at UHF and microwave frequencies.

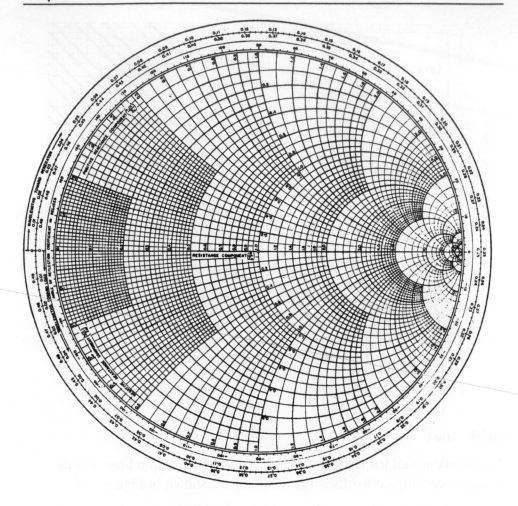

Figure 3-9: Smith Chart

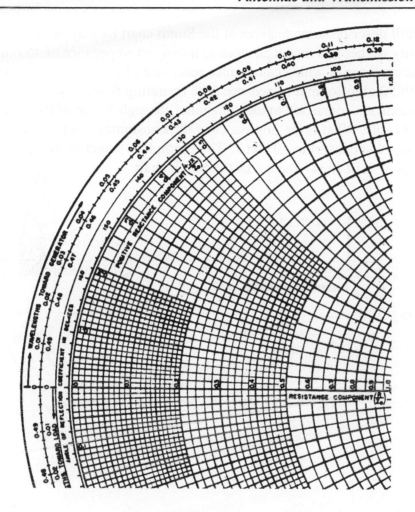

Figure 3-10: Expanded View of Smith Chart

We'll describe some features of the Smith chart by way of an example. Let's say we need to match an antenna having an impedance of 15 ohms resistance in series with a capacitance reactance of 75 ohms to a transmitter with 50 ohms output impedance. The operating frequency is 173 MHz. The antenna is connected to the transmitter through 73 cm of RG-58C coaxial cable. What is the impedance at the transmitter that the matching network must convert to 50 ohms? The example is sketched in Figure 3-11 and Figure 3-12 shows the use of the Smith chart.

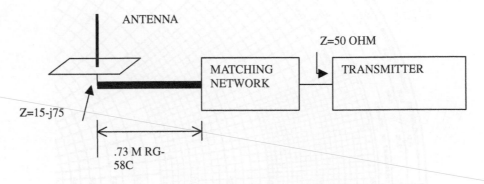

Figure 3-11: Antenna Matching Example

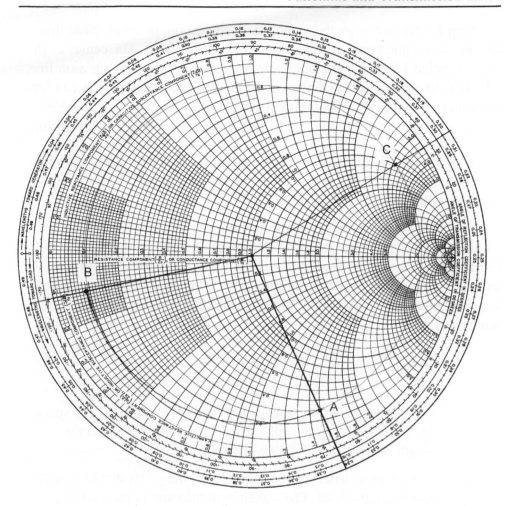

Figure 3-12: Using the Smith Chart

Step 1. First we mark the antenna impedance on the chart. Note that the resistance and reactance coordinates are normalized. The center of the chart, labeled 1.0, is the characteristic impedance of the transmission line, which is 50 ohms. We divide the resistance and capacitive reactance of the antenna by 50 and get, in complex form:

$$Z_{load} = 0.3 - j1.5$$

This is marked at the intersection of the 0.3 resistance circle with the 1.5 capacitive reactance coordinate in the bottom half of the chart. This point is marked "A" in Figure 3-12.

Step 2. The impedance at the transmitter end of the transmission line is located on a circle whose radius is the length of a line from the center of the chart to point "A" (assuming no cable losses). In order to find the exact location of the impedance on this circle for the 73-cm coax cable, we must relate the physical cable length, l, to the electrical length, L, in wavelengths.

$$L = \frac{l}{\eta\lambda}$$

where η is the velocity factor (.66) and λ is the wavelength in free space (3×10^8/frequency). Inserting the values for this example we find the electrical length of the line is 0.64 wavelengths.

The Smith chart instructs us to move toward the generator in a clockwise direction from the load. The wavelength measure is marked on the outmost marked circle. Every 0.5 wavelengths, the load impedance is reflected to the end of the cable with no change, so we subtract a whole number of half wavelengths, in this case one, from the cable length, giving us

$$.64\lambda_c - .5\,\lambda_c = 0.14\lambda_c$$

where λ_c is the wavelength in the cable.

The line drawn on Figure 3-12 from the center through Z_{load} (point A) intersects the "wavelengths to generator" circle at 0.342. We add 0.14 to get 0.482 and draw a line from the center to this point. Mark the line at point B, which is the same distance from the center as point A. This is done conveniently with a compass.

Step 3. Read off point B. It is $0.1 - j0.113$. Multiplying by the 50-ohm characteristic impedance we get

$$Z_{gen} = 5 - j5.65 \text{ ohms.}$$

Step 4. Now using the procedure of Example 3 above, we can design a matching network to match the impedance seen at the coax cable to the 50-ohm impedance of the transmitter. You have to add the capacitive reactance, 5.65 ohms, to the reactance you find for L_1. The resulting network components are $L_1 = 19$ nH and $C_1 = 55$ pF.

By examining the Smith chart in Figure 3-12, we note that if we use a longer coaxial cable we can get an impedance at its end that has a real part, or resistance, of 50 ohms, and an inductive reactance of $50 \times 3.1 = 155$ ohms. This is point C in the figure. We attain this impedance by adding $.22\lambda_c = 25$ cm to the original transmission line, for a total coax cable length of 98 cm. Now the only matching component we need is a series capacitor to cancel out the inductive reaction of 155 ohms. Using the "Conversions" worksheet we find its capacitance to be approximately 6 pF.

Other transmission-line matching problems can be solved using the Smith chart. Using the chart, you can easily determine *SWR*, reflection coefficient and return loss. The chart also has provision for accounting for line losses.

The Smith chart is very handy for seeing at a glance the effects on impedance of changing transmission line lengths, and also for using transmission lines as matching networks. Computer programs are available for doing Smith chart plotting. The enclosed Mathcad worksheet "Transmission Lines" solves transmission line problems directly from mathematical formulas.

Microstrip

From around 800 MHz and higher, the lengths of printed circuit board conductors are a significant fraction of a wavelength, so they act as transmission lines. Thus, if the input to a receiver integrated circuit or low-noise amplifier is 50 ohms, and a conductor length of 6 cm connects to the antenna socket, the RF plug from the antenna will *not* see 50 ohms, unless the conductor is designed to have a characteristic impedance of 50 ohms. A printed conductor over a ground plane (copper plating on the opposite side of the board) is called microstrip. The transmission line characteristics of conductors on a board are used in UHF and microwave circuits as matching networks between the various components.

Using the attached Mathcad worksheet "Microstrip," you can find the conductor width required to get a required characteristic line impedance, or you can find the impedance if you know the width. Then you can use the Smith chart to do impedance transformations and design matching networks using microstrip components. The "Microstrip" worksheet also gives you the wavelength on the pc board for a given frequency. In order to use this worksheet, you have to know the dielectric constant of your board material and the board's thickness.

3.5 Measuring Techniques

If you happen to have a vector analyzer, you can measure the impedances you want to match, design a matching network, and check the accuracy of your design. When a matching network is designed and adjusted correctly, the impedance looking into the network where it is connected to the load or source is the complex conjugate of the impedance of the load or of the source impedance. The complex conjugate of an impedance has the same real part as the impedance and minus the imaginary part of the impedance. For example, if $Z_{source} = 30 - j12$ ohms, then the impedance seen at the input to the matching network should be $30 + j12$ ohms when its output port is connected to the load.

Without a vector analyzer, you need considerable cut-and-try to optimize the antenna and matching components. Other instruments, usually available in RF electronics laboratories, can be a big help. Here are some ideas for adjusting antennas and circuits for resonance at the operating frequency using relatively inexpensive equipment (compared to a vector analyzer).

A grid dip meter (still called that, although for years it has been based on a transistor oscillator, not a vacuum tube) is a simple, inexpensive tool, popular with radio amateurs. It consists of a tunable RF oscillator with external coil, allowing it to be lightly coupled to a resonant circuit, which can be an antenna of almost any type. When the dip meter is tuned across the resonant frequency of the passive circuit under test, its indicating meter shows a current dip due to absorption of energy from the instrument's oscillator. A loop antenna with resonating capacitor is easy to adjust using this method. A dipole, ground plane, or helical antenna can also be checked for resonance by connecting a small one-turn loop to the antenna terminals with matching circuit components disconnected. Set the dip meter coil close to the loop and tune the instrument to find a dip.

The main limitation to the grid dip meter is its frequency range, usually no more than 250 MHz. Higher frequency resonances can be measured with a return loss bridge, also called directional bridge or impedance bridge. This device, which is an integral part of a scalar network analyzer, can be used with a spectrum analyzer and tracking generator or a noise source to give a relative display of return loss versus frequency.

The return loss bridge is a three-port device that indicates power reflected from a mismatched load. Figure 3-13 shows a diagram of the bridge. Power applied at the source port passes to the device under test at the test port with a nominal attenuation of 6 dB. Power reflected from the device under test appears at the measurement port, attenuated approximately 6 dB. If the tested circuit presents the same impedance as the characteristic impedance of the bridge, there will be no output at the measurement port, except for a leakage output on the order of 50 dB below the output of the source. If the test port sees an open or short circuit, all power will be reflected and the measurement port output will be around −12 dB. The return loss is the difference between the output measured in dBm at the measurement port when the test port is open or shorted, and the dBm output when the circuit under test is connected to the test port.

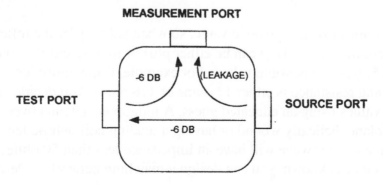

Figure 3-13: Return Loss Bridge

A setup to determine the resonant frequency of an antenna is shown in Figure 3-14 (a). The antenna is connected to the test terminal of the bridge through a short length of 50-ohm coaxial cable. A spectrum analyzer is connected to the measurement port and a tracking generator, whose frequency is swept in tandem with the frequency sweep of the spectrum analyzer, drives the source port.

When the swept frequency passes the resonant frequency of the antenna, the analyzer display dips at that frequency. At the resonant frequency, reactance is cancelled and the antenna presents a pure resistance. The closer the antenna impedance is to 50 ohms, the deeper the dip. Antenna parameters may be changed—length, loading coil or helical coil dimensions, for example—until the dip occurs at the desired operating frequency. Dips are usually observed at several frequencies because of more than one resonance in the system. By noting the effect of changes in the antenna on the various dips, as well as designing the antenna properly in the first place to give approximately the correct resonant frequency, the right dip can usually be correctly identified.

You can get an approximation of the resonant antenna resistance R_{ant} by measuring the return loss RL and then converting it to resistance using the following equations, derived from Equations (3-11) and (3-13), or by using the Mathcad "Transmission Lines" worksheet.

$$\rho = \pm 10^{\frac{-RL}{20}} \qquad (3\text{-}14)$$

$$R_{ant} = Z_0 \cdot \frac{1+\rho}{1-\rho} \qquad (3\text{-}15)$$

The return loss is a positive value, so when solving for the reflection coefficient in Eq. (3-14), ρ can be either plus or minus, and R_{ant} found in Eq. (3-15) has two possible values. For example, if the return loss is 5 dB, the antenna resistance is either 14 ohms or 178 ohms. You decide between the two values using an educated guess. A monopole antenna over a ground plane, helically wound or having a loading coil, whose length is less than a quarter wave will have an impedance less than 50 ohms. Once the resistance is known, you can design a matching network as described above.

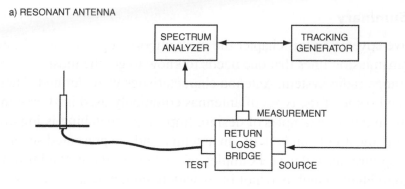

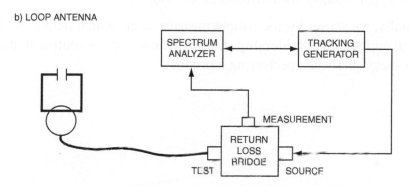

Figure 3-14: Resonant Circuit Test Setup

The arrangement shown in Figure 3-14(b) is convenient for checking the resonant frequency of a loop antenna, up to around 500 MHz. Use a short piece of coax cable and a loop of stiff magnet wire with a diameter of 2 cm. Use two or three turns in the loop for VHF and lower frequencies. At loop resonance, the spectrum analyzer display shows a sharp dip. Keep the test coil as far as possible from the loop, while still seeing the dip, to avoid influencing the circuit. You can easily tune the loop circuit, if it has a trimmer capacitor, by observing the location of the dip. The same setup can be used for checking resonance of tuning coils in the transmitter or receiver. There must be no radiation from the circuit when this test is made. If possible, disable the oscillator and apply power to the device being tested. The resonant frequency of a tuned circuit that is coupled to a transistor stage will be different when voltage is applied and when it is not.

3.6 Summary

We have covered in this chapter the most important properties of antennas and transmission lines that one needs to know to get the most from a short-range radio system. Antenna characteristics were defined. Then we discussed some of the types of antennas commonly used in short-range systems and gave examples of design. Impedance matching is imperative to get the most into, and out of, an antenna, and we presented several matching circuits and gave examples of how to use them. We introduced the Smith chart, which may not be as widely used now as it once was as a design tool, but understanding it helps us visualize the concepts of circuit matching, particularly with distributed components.

Finally, we showed some simple measurements which help in realizing a design and which considerably shorten the cut-and-try routine that is almost inevitable when perfecting a product.

CHAPTER 4

Communication Protocols and Modulation

In this chapter we take an overall view of the characteristics of the communication system. While these characteristics are common to any wireless communication link, for detail we'll address the peculiarities of short-range systems.

A simple block diagram of a digital wireless link is shown in Figure 4-1. The link transfers information originating at one location, referred to as source data, to another location where it is referred to as reconstructed data. A more concrete implementation of a wireless system, a security system, is shown in Figure 4-2.

4.1 Baseband Data Format and Protocol

Let's first take a look at what information we may want to transfer to the other side. This is important in determining what bandwidth the system needs.

Change-of-state source data

Many short-range systems only have to relay information about the state of a contact. This is true of the security system of Figure 4-2 where an infrared motion detector notifies the control panel when motion is detected. Another example is the push-button transmitter, which may be used as a panic button or as a way to activate and deactivate the control system, or a wireless smoke detector, which gives advance warning of an impending fire. There are also what are often referred to as "technical" alarms—gas detectors, water level detectors, and low and high temperature detectors—whose function is to give notice of an abnormal situation.

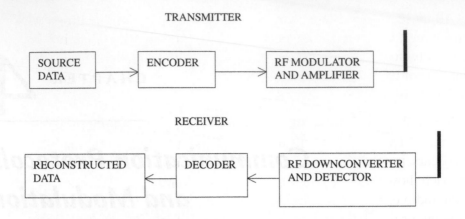

Figure 4-1: Radio Communication Link Diagram

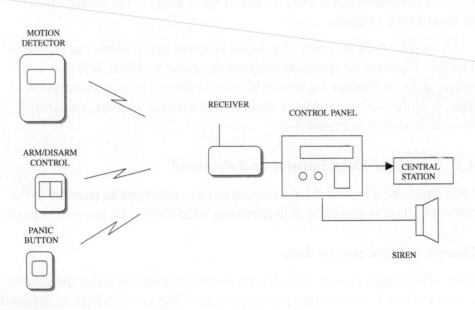

Figure 4-2: Security System

All these examples are characterized as very low-bandwidth information sources. Change of state occurs relatively rarely, and when it does, we usually don't care if knowledge of the event is signaled tens or even hundreds of milliseconds after it occurs. Thus, required information bandwidth is very low—several hertz.

It would be possible to maintain this very low bandwidth by using the source data to turn on and off the transmitter at the same rate the information occurs, making a very simple communication link. This is not a practical approach, however, since the receiver could easily mistake random noise on the radio channel for a legitimate signal and thereby announce an intrusion, or a fire, when none occurred. Such false alarms are highly undesirable, so the simple on/off information of the transmitter must be coded to be sure it can't be misinterpreted at the receiver.

This is the purpose of the encoder shown in Figure 4-1. This block creates a group of bits, assembled into a frame, to make sure the receiver will not mistake a false occurrence for a real one. Figure 4-3 is an example of a message frame. The example has four fields. The first field is a preamble with start bit, which conditions the receiver for the transfer of information and tells it when the message begins. The next field is an identifying address. This address is unique to the transmitter and its purpose is to notify the receiver from where, or from what unit, the message is coming. The data field follows, which may indicate what type of event is being signaled, followed, in some protocols, by a parity bit or bits to allow the receiver to determine whether the message was received correctly.

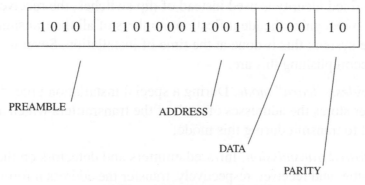

Figure 4-3: Message Frame

Address Field

The number of bits in the address field depends on the number of different transmitters there may be in the system. Often the number of possibilities is far greater than this, to prevent confusion with neighboring, independent systems and to prevent the statistically possible chance that random noise will duplicate the address. The number of possible addresses in the code is 2^{L1}, where $L1$ is the length of the message field. In many simple security systems the address field is determined by dip switches set by the user. Commonly, eight to ten dip switch positions are available, giving 256 to 1024 address possibilities. In other systems, the address field, or device identity number, is a code number set in the unit micro-controller during manufacture. This code number is longer than that produced by dip switches, and may be 16 to 24 bits long, having 65,536 to 16,777,216 different codes. The longer codes greatly reduce the chances that a neighboring system or random event will cause a false alarm. On the other hand, the probability of detection is lower with the longer code because of the higher probability of error. This means that a larger signal-to-noise ratio is required for a given probability of detection.

In all cases, the receiver must be set up to recognize transmitters in its own system. In the case of dip-switch addressing, a dip switch in the receiver is set to the same address as in the transmitter. When several transmitters are used with the same receiver, all transmitters must have the same identification address as that set in the receiver. In order for each individual transmitter to be recognized, a subfield of two to four extra dip switch positions can be used for this differentiation. When a built-in individual fixed identity is used instead of dip switches, the receiver must be taught to recognize the identification numbers of all the transmitters used in the system; this is done at the time of installation. Several common ways of accomplishing this are:

(a) *Wireless "learn" mode.* During a special installation procedure, the receiver stores the addresses of each of the transmitters which are caused to transmit during this mode;

(b) *Infrared transmission.* Infrared emitters and detectors on the transmitter and receiver, respectively, transfer the address information;

(c) *Direct key-in*. Each transmitter is labeled with its individual address, which is then keyed into the receiver or control panel by the system installer;

(d) *Wired learn mode*. A short cable temporarily connected between the receiver and transmitter is used when performing the initial address recognition procedure during installation.

Advantages and disadvantages of the two addressing systems

Dip switch

Advantages	Disadvantages
Unlimited number of transmitters can be used with a receiver.	Limited number of bits increases false alarms and interference from adjacent systems.
Can be used with commercially available data encoders and decoders.	Device must be opened for coding during installation.
Transmitter or receiver can be easily replaced without recoding the opposite terminal.	Multiple devices in a system are not distinguishable in most simple systems.
	Control systems are vulnerable to unauthorized operation since the address code can be duplicated by trial and error.

Internal fixed code identity

Advantages	Disadvantages
Large number of code bits reduces possibility of false alarms.	Longer code reduces probability of detection.
System can be set up without opening transmitter.	Replacing transmitter or receiver involves redoing the code learning procedure.
Each transmitter is individually recognized by receiver.	Limited number of tranmitters can be used with each receiver.
	Must be used with a dedicated microcontroller. Cannot be used with standard encoders and decoders.

Code-hopping addressing

While using a large number of bits in the address field reduces the possibility of false identification of a signal, there is still a chance of purposeful duplication of a transmitter code to gain access to a controlled entry. Wireless push buttons are used widely for access control to vehicles and buildings. Radio receivers exist, popularly called "code grabbers," which receive the transmitted entry signals and allow retransmitting them for fraudulent access to a protected vehicle or other site. To counter this possibility, addressing techniques were developed that cause the code to change every time the push button is pressed, so that even if the transmission is intercepted and recorded, its repetition by a would-be intruder will not activate the receiver, which is now expecting a different code. This method is variously called code rotation, code hopping, or rolling code addressing. In order to make it virtually impossible for a would-be intruder to guess or try various combinations to arrive at the correct code, a relatively large number of address bits are used. In some devices, 36-bit addresses are employed, giving a total of over 68 billion possible codes.

In order for the system to work, the transmitter and receiver must be synchronized. That is, once the receiver has accepted a particular transmission, it must know what the next transmitted address will be. The addresses cannot be sequential, since that would make it too easy for the intruder to break the system. Also, it is possible that the user might press the push button to make a transmission but the receiver may not receive it, due to interference or the fact that the transmitter is too far away. This could even happen several times, further unsynchronizing the transmitter and the receiver. All of the code-hopping systems are designed to prevent such unsynchronization.

Following is a simplified description of how code hopping works, aided by Figure 4-4.

Both the receiver and the transmitter use a common algorithm to generate a pseudorandom sequence of addresses. This algorithm works by manipulating the address bits in a certain fashion. Thus, starting at a known address, both sides of the link will create the same next address. For demonstration purposes, Figure 4-4 shows the same sequence of two-digit decimal numbers at the transmitting side and the receiving side. The solid transmitter arrow points to the present transmitter address and the solid receiver arrow points to the expected receiver address. After trans-

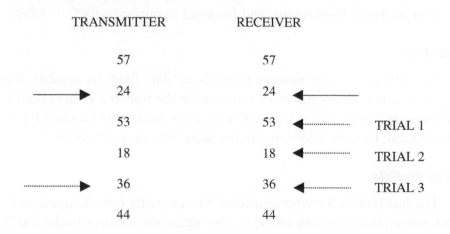

TRANSMITTER	RECEIVER	
57	57	
24	24	
53	53	TRIAL 1
18	18	TRIAL 2
36	36	TRIAL 3
44	44	

Figure 4-4: Code Hopping

mission and reception, both transmitter and receiver calculate their next addresses, which will be the same. The arrows are synchronized to point to the same address during a system set-up procedure. As long as the receiver doesn't miss a transmission, there is no problem, since each side will calculate an identical next address. However, if one or more transmissions are missed by the receiver, when it finally does receive a message, its expected address will not match the received address. In this case it will perform its algorithm again to create a new address and will try to match it. If the addresses still don't match, a new address is calculated until either the addresses match or a given number of trials have been made with no success. At this point, the transmitter and receiver are unsynchronized and the original setup procedure has to be repeated to realign the transmitter and receiver addresses.

The number of trials permitted by the receiver may typically be between 64 and 256. If this number is too high, the possibility of compromising the system is greater (although with a 36-bit address a very large number of trials would be needed for this) and with too few trials, the frequency of inconvenient resynchronization would be greater. Note that a large number of trials takes a lot of time for computations and may cause a significant delay in response.

Several companies make rolling code components, among them Microchip, Texas Instruments, and National Semiconductor.

Data Field

The next part of the message frame is the data field. Its number of bits depends on how many pieces of information the transmitter may send to the receiver. For example, the motion detector may transmit three types of information: motion detection, tamper detection, or low battery.

Parity Bit Field

The last field is for error detection bits, or parity bits. As discussed later, some protocols have inherent error detection features so the last field is not needed.

Baseband Data Rate

Once we have determined the data frame, we can decide on the appropriate baseband data rate. For the security system example, this rate will usually be several hundred hertz up to a maximum of a couple of kilohertz. Since a rapid response is not needed, a frame can be repeated several times to be more certain it will get through. Frame repetition is needed in systems where space diversity is used in the receiver. In these systems, two separate antennas are periodically switched to improve the probability of reception. If signal nulling occurs at one antenna because of the multipath phenomena, the other antenna will produce a stronger signal, which can be correctly decoded. Thus, a message frame must be sent more often to give it a chance to be received after unsuccessful reception by one of the antennas.

Supervision

Another characteristic of digital event systems is the need for link supervision. Security systems and other event systems, including medical emergency systems, are one-way links. They consist of several transmitters and one receiver. As mentioned above, these systems transmit relatively rarely, only when there is an alarm or possibly a low-battery condition. If a transmitter ceases to operate, due to a component failure, for example, or if there is an abnormal continuing interference on the radio channel, the fact that the link has been broken will go undetected. In the case of a security system, the installation will be unprotected, possibly, until a routine system inspection is carried out. In a wired system, such a possibility is usually covered by a normally energized relay connected through closed contacts to a control panel. If a fault occurs in the device, the relay becomes unenergized and the panel detects the opening of the contacts. Similarly, cutting the connecting wires will also be detected by the panel. Thus, the advantages of a wireless system are compromised by the lower confidence level accompanying its operation.

Many security systems minimize the risk of undetected transmitter failure by sending a supervisory signal to the receiver at a regular interval. The receiver expects to receive a signal during this interval and can emit a supervisory alarm if the signal is not received. The supervisory signal must be identified as such by the receiver so as not to be mistaken for an alarm.

The duration of the supervisory interval is determined by several factors:

■ Devices certified under FCC Part 15 paragraph 15.231, which applies to most wireless security devices in North America, may not send regular transmissions more frequently than one per hour.

■ The more frequently regular supervision transmissions are made, the shorter the battery life of the device.

■ Frequent supervisory transmissions when there are many transmitters in the system raise the probability of a collision with an alarm signal, which may cause the alarm not to get through to the receiver.

■ The more frequent the supervisory transmissions, the higher the confidence level of the system.

While it is advantageous to notify the system operator at the earliest sign of transmitter malfunction, frequent supervision raises the possibility that a fault might be reported when it doesn't exist. Thus, most security systems determine that a number of consecutive missing supervisory transmissions must be detected before an alarm is given. A system which specifies security emissions once every hour, for example, may wait for eight missing supervisory transmissions, or eight hours, before a supervisory alarm is announced. Clearly, the greater the consequences of lack of alarm detection due to a transmitter failure, the shorter the supervision interval must be.

Continuous digital data

In other systems flowing digital data must be transmitted in real time and the original source data rate will determine the baseband data rate. This is the case in wireless LANs and wireless peripheral-connecting devices. The data is arranged in message frames, which contain fields needed for correct transportation of the data from one side to the other, in addition to the data itself.

An example of a frame used in *synchronous data link control* (SDLC) is shown in Figure 4-5. It consists of beginning and ending bytes that delimit the frame in the message, address and control fields, a data field of undefined length, and check bits or parity bits for letting the receiver check whether the frame was correctly received. If it is, the receiver sends a short acknowledgment and the transmitter can continue with the next frame. If no acknowledgment is received, the transmitter repeats the message again and again until it is received. This is called an ARQ (automatic repeat query) protocol. In high-noise environments, such as encountered on radio channels, the repeated transmissions can significantly slow down the message throughput.

More common today is to use a *forward error control* (FEC) protocol. In this case, there is enough information in the parity bits to allow the receiver to correct a small number of errors in the message so that it will not have to request retransmission. Although more parity bits are needed for error correction than for error detection alone, the throughput is greatly increased when using FEC on noisy channels.

In all cases, we see that extra bits must be included in a message to insure proper transmission, and the consequently longer frames require a higher transmission rate than what would be needed for the source data alone. This message overhead must be considered in determining the required bit rate on the channel, the type of digital modulation, and consequently the bandwidth.

BEGINNING FLAG - 8 BITS	ADDRESS - 8 BITS	CONTROL - 8 BITS	INFORMATION - ANY NO. OF BITS	ERROR DETECTION - 16 BITS	ENDING FLAG - 8 BITS

Figure 4-5: Synchronous Data Link Control Frame

Analog transmission

Analog transmission devices, such as wireless microphones, also have a baseband bandwidth determined by the data source. A high-quality wireless microphone may be required to pass 50 to 15,000 Hz, whereas an analog wireless telephone needs only 100 to 3000 Hz. In this case determining the channel bandwidth is more straightforward than in the digital case, although the bandwidth depends on whether AM or FM modulation is used. In most short-range radio applications, FM is preferred—narrowband FM for voice communications and wide-band FM for quality voice and music transmission.

4.2 Baseband Coding

The form of the information signal that is modulated onto the RF carrier we call here baseband coding. We refer below to both digital and analog systems, although strictly speaking the analog signal is not coded but is modified to obtain desired system characteristics.

Digital systems

Once we have a message frame, composed as we have shown by address and data fields, we must form the information into signal levels that can be effectively transmitted, received, and decoded. Since we're concerned here with binary data transmission, the baseband coding selected has to give the signal the best chance to be decoded after it has been modified by noise and the response of the circuit and channel elements. This coding consists essentially of the way that zeros and ones are represented in the signal sent to the modulator of the transmitter.

There are many different recognized systems of baseband coding. We will examine only a few common examples.

These are the dominant criteria for choosing or judging a baseband code:

(a) *Timing*. The receiver must be able to take a data stream polluted by noise and recognize transitions between each bit. The bit transitions must be independent of the message content—that is, they must be identifiable even for long strings of zeros or ones.

(b) *DC content.* It is desirable that the average level of the message—that is, its DC level—remains constant throughout the message frame, regardless of the content of the message. If this is not the case, the receiver detection circuits must have a frequency response down to DC so that the levels of the message bits won't tend to wander throughout the frame. In circuits where coupling capacitors are used, such a response is impossible.

(c) *Power spectrum.* Baseband coding systems have different frequency responses. A system with a narrow frequency response can be filtered more effectively to reduce noise before detection.

(d) *Inherent error detection.* Codes that allow the receiver to recognize an error on a bit-by-bit basis have a lower possibility of reporting a false alarm when error-detecting bits are not used.

(e) *Probability of error.* Codes differ in their ability to properly decode a signal, given a fixed transmitter power. This quality can also be stated as having a lower probability of error for a given signal-to-noise ratio.

(f) *Polarity independence.* There is sometimes an advantage in using a code that retains its characteristics and decoding capabilities when inverted. Certain types of modulation and demodulation do not retain polarity information. Phase modulation is an example.

Now let's look at some common codes (Figure 4-6) and rate them according to the criteria above. It should be noted that coding considerations for event-type reporting are far different from those for flowing real-time data, since bit or symbol times are not so critical, and message frames can be repeated for redundancy to improve the probability of detection and reduce false alarms. In data flow messages, the data rate is important and sophisticated error detection and correction techniques are used to improve system sensitivity and reliability.

a) *Non-return to Zero (NRZ)—Figure 4-6a.*

This is the most familiar code, since it is used in digital circuitry and serial wired short-distance communication links, like RS-232. However, it is rarely used directly for wireless communication. Strings of ones or zeros leave it without defined bit boundaries, and its DC level is very dependent on the message content. There is no inherent error detection. If

NRZ coding is used, an error detection or correction field is imperative. If amplitude shift keying (ASK) modulation is used, a string of zeros means an extended period of no transmission at all. In any case, if NRZ signaling is used, it should only be for very short frames of no more than eight bits.

(b) *Manchester code—Figure 4-6b*.

A primary advantage of this code is its relatively low probability of error compared to other codes. It is the code used in Ethernet local area networks. It gives good timing information since there is always a transition in the middle of a bit, which is decoded as zero if this is a positive transition and a one otherwise. The Manchester code has a constant DC component and its waveform doesn't change if it passes through a capacitor or transformer. However, a "training" code sequence should be inserted before the message information as a preamble to allow capacitors in the receiver detection circuit to reach charge equilibrium before the actual message bits appear. Inverting the Manchester code turns zeros to ones and ones to zeros. The frequency response of Manchester code has components twice as high as NRZ code, so a low-pass filter in the receiver must have a cut-off frequency twice as high as for NRZ code with the same bit rate.

(c) *Biphase Mark—Figure 4-6c*.

This code is somewhat similar to the Manchester code, but bit identity is determined by whether or not there is a transition in the middle of a bit. For biphase mark, a level transition in the middle of a bit (going in either direction) signifies a one, and a lack of transition indicates zero. Biphase space is also used, where the space character has a level transition. There is always a transition at the bit boundaries, so timing content is good. A lack of this transition gives immediate notice of a bit error and the frame should then be aborted. The biphase code has constant DC level, no matter what the message content, and a preamble should be sent to allow capacitor charge equalization before the message bits arrive. As with the Manchester code, frequency content is twice as much as for the NRZ code. The biphase mark or space code has the added advantage of being polarity independent.

(d) *Pulse width modulation—Figure 4-6d.*

As shown in the figure, a one has two timing durations and a zero has a pulse width of one duration. The signal level inverts with each bit so timing information for synchronization is good. There is a constant average DC level. Since the average pulse width varies with the message content, in contrast with the other examples, the bit rate is not constant. This code has inherent error detection capability.

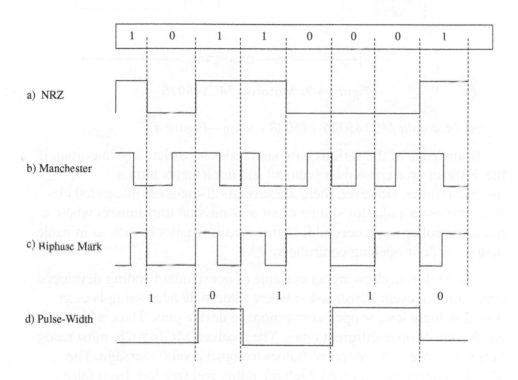

Figure 4-6: Baseband Bit Formats

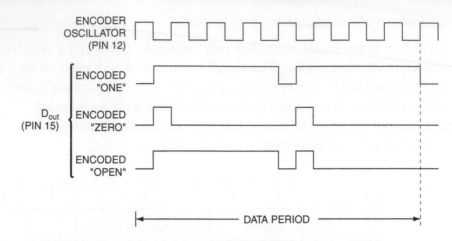

Figure 4-7: Motorola MC145026

(e) *Motorola MC145026-145028 coding—Figure 4-7.*

Knowledge of the various baseband codes is particularly important if the designer creates his own protocol and implements it on a microcontroller. However, there are several off-the-shelf integrated circuits that are popular for simple event transmission transmitters where a microcontroller is not needed for other circuit functions, such as in panic buttons or door-opening controllers.

The Motorola chips are an example of nonstandard coding developed especially for event transmission where three-state addressing is determined as high, low, or open connections to device pins. Thus, a bit symbol can be one of three different types. The receiver, MC145028, must recognize two consecutive identical frames to signal a valid message. The Motorola protocol gives very high reliability and freedom from false alarms. Its signal does have a broad frequency spectrum relative to the data rate and the receiver filter passband must be designed accordingly. The DC level is dependent on the message content.

Analog baseband conditioning

Wireless microphones and headsets are examples of short-range systems that must maintain high audio quality over the vagaries of changing path lengths and indoor environments, while having small size and low cost. To help them achieve this, they have a signal conditioning element in their baseband path before modulation. Two features used to achieve high signal-to-noise ratio over a wide dynamic range are pre-emphasis/de-emphasis and compression/expansion. Their positions in the transmitter/receiver chain are shown in Figure 4-8.

The transmitter audio signal is applied to a high-pass filter (pre-emphasis) which increases the high frequency content of the signal. In the receiver, the detected audio goes through a complementary low-pass filter (de-emphasis), restoring the signal to its original spectrum composition. However, in so doing, high frequency noise that entered the signal path after the modulation process is filtered out by the receiver while the desired signal is returned to its original quality.

Compression of the transmitted signal raises the weak sounds and suppresses strong sounds to make the modulation more efficient. Reversing the process in the receiver weakens annoying background noises while restoring the signal to its original dynamic range.

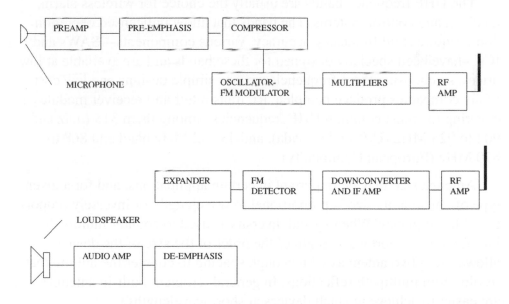

Figure 4-8: Wireless Microphone System

4.3 RF Frequency and Bandwidth

There are several important factors to consider when determining the radio frequency of a short-range system:

- Telecommunication regulations

- Antenna size

- Cost

- Interference

- Propagation characteristics

When you want to market a device in several countries and regions of the world, you may want to choose a frequency that can be used in the different regions, or at least frequencies that don't differ very much so that the basic design won't be changed by changing frequencies. We'll discuss the matter of frequency choice and regulations in detail in Chapter 9.

The UHF frequency bands are usually the choice for wireless alarm, medical, and control systems. The bands that allow unlicensed operation don't require rigid frequency accuracy. Various components—SAWs and ICs—have been specially designed for these bands and are available at low prices, so choosing these frequencies means simple designs and low cost. Many companies produce complete RF transmitter and receiver modules covering the most common UHF frequencies, among them 315 MHz and 902 to 928 MHz (U.S. and Canada), and 433.92 MHz band and 868 to 870 MHz (European Community).

Antenna size may be important in certain applications, and for a given type of antenna, its size is proportional to wavelength, or inversely proportional to frequency. When spatial diversity is used to counter multipath interference, a short wavelength of the order of the size of the device allows using two antennas with enough spacing to counter the nulling that results from multipath reflections. In general, efficient built-in antennas are easier to achieve in small devices at short wavelengths.

From VHF frequencies and up, cost is directly proportional to increased frequency.

Natural and manmade background noise is higher on the lower frequencies. On the other hand, certain frequency bands available for short-range use may be very congested with other users, such as the ISM bands. Where possible, it is advisable to chose a band set aside for a particular use, such as the 868 – 870 MHz band available in Europe.

Propagation characteristics also must be considered in choosing the operating frequency. High frequencies reflect easily from surfaces but penetrate insulators less readily than lower frequencies.

The radio frequency bandwidth is a function of the baseband bandwidth and the type of modulation employed. For security event transmitters, the required bandwidth is small, of the order of several kilohertz. If the complete communication system were designed to take advantage of this narrow bandwidth, there would be significant performance advantages over the most commonly used systems having a bandwidth of hundreds of kilohertz. For given radiated transmitter power, the range is inversely dependent on the receiver bandwidth. Also, narrow-band unlicensed frequency allotments can be used where available in the different regions, reducing interference from other users. However, cost and complexity considerations tend to outweigh communication reliability for these systems, and manufacturers decide to make do with the necessary performance compromises. The bandwidth of the mass production security devices is thus determined by the frequency stability of the transmitter and receiver frequency determining elements, and not by the required signaling bandwidth. The commonly used SAW devices dictate a bandwidth of at least 200 kHz, whereas the signaling bandwidth may be only 2 kHz. Designing the receiver with a passband of 20 kHz instead of 200 kHz would increase sensitivity by 10 dB, roughly doubling the range. This entails using stable crystal oscillators in the transmitter and in the receiver local oscillator.

4.4 Modulation

Amplitude modulation (AM) and frequency modulation (FM), known from commercial broadcasting, have their counterparts in modulation of digital signals, but you must be careful before drawing similar conclusions about the merits of each. The third class of modulation is phase modulation, not used in broadcasting, but its digital counterpart is commonly used in high-end, high-data-rate digital wireless communication.

Digital AM is referred to as ASK—amplitude shift keying—and sometimes as OOK—on/off keying. FSK is frequency shift keying, the parallel to FM. Commercial AM has a bandwidth of 10 kHz whereas an FM broadcasting signal occupies 180 kHz. The high post-detection signal-to-noise ratio of FM is due to this wide bandwidth. However, on the negative side, FM has what is called a threshold effect, also due to wide bandwidth. Weak FM signals are unintelligible at a level that would still be usable for AM signals. When FM is used for two-way analog communication, narrow-band FM, which occupies a similar bandwidth to AM, also has comparable sensitivity for a given S/N.

Modulation for digital event communication

For short-range digital communication we're not interested in high fidelity, but rather high sensitivity. Other factors for consideration are simplicity and cost of modulation and demodulation. Let's now look into the reasons for choosing one form of modulation or the other.

An analysis of error rates versus bit energy to noise density shows that there is no inherent advantage of one system, ASK or FSK, over the other. This conclusion is based on certain theoretical assumptions concerning bandwidth and method of detection. While practical implementation methods may favor one system over the other, we shouldn't jump to conclusions that FSK is necessarily the best, based on a false analogy to FM and AM broadcasting.

In low-cost security systems, ASK is the simplest and cheapest method to use. For this type of modulation we must just turn on and turn off the radio frequency output in accordance with the digital modulating signal. The output of a microcontroller or dedicated coding device biases on and off a single SAW-controlled transistor RF oscillator. Detection in the receiver is also simple. It may be accomplished by a diode detector in several receiver architectures, to be discussed later, or by the RSSI (received signal strength indicator) output of many superheterodyne receiver ICs employed today. Also, ASK must be used in the still widespread superregenerative receivers.

For FSK, on the other hand, it's necessary to shift the transmitting frequency between two different values in response to the digital code.

More elaborate means is needed for this than in the simple ASK transmitter, particularly when crystal or SAW devices are used to keep the frequency stable. In the receiver, also, additional components are required for FSK demodulation as compared to ASK. We have to decide whether the additional cost and complexity is worthwhile for FSK.

In judging two systems of modulation, we must base our results on a common parameter that is a basis for comparison. This may be peak or average power. For FSK, the peak and average powers are the same. For ASK, average power for a given peak power depends on the duty cycle of the modulating signal. Let's assume first that both methods, ASK and FSK, give the same performance—that is, the same sensitivity—if the average power in both cases are equal. It turns out that in this case, our preference depends on whether we are primarily marketing our system in North America or in Europe. This is because of the difference in the definition of the power output limits between the telecommunication regulations in force in the US and Canada as compared to the common European regulations.

The US FCC Part 15 and similar Canadian regulations specify an *average* field strength limit. Thus, if the transmitter is capable of using a peak power proportional to the inverse of its modulation duty cycle, while maintaining the allowed average power, then under our presumption of equal performance for equal average power, there would be no reason to prefer FSK, with its additional complexity and cost, over ASK.

In Western Europe, on the other hand, the low-power radio specification, ETSI 300 220 limits the *peak* power of the transmitter. This means that if we take advantage of the maximum allowed peak power, FSK is the proper choice, since for a given peak power, the average power of the ASK transmitter will always be less, in proportion to the modulating signal duty cycle, than that of the FSK transmitter.

However, is our presumption of equal performance for equal average power correct? Under conditions of added white Gaussian noise (AWGN) it seems that it is. This type of noise is usually used in performance calculations since it represents the noise present in all electrical circuits as well as cosmic background noise on the radio channel. But in real life, other forms of interference are present in the receiver passband that have very different, and usually unknown, statistical characteristics from AWGN. On the UHF frequencies normally used for short-range radio, this interference is primarily from other transmitters using the same or nearby frequencies. To compare performance, we must examine how the ASK and FSK receivers handle this type of interference. This examination is pertinent in the US and Canada where we must choose between ASK and FSK when considering that the average power, or signal-to-noise ratio, remains constant. Some designers believe that a small duty cycle resulting in high peak power per bit is advantageous since the presence of the bit, or high peak signal, will get through a background of interfering signals better than another signal with the same average power but a lower peak. To check this out, we must assume a fair and equal basis of comparison. For a given data rate, the low-duty-cycle ASK signal will have shorter pulses than for the FSK case. Shorter pulses means higher baseband bandwidth and a higher cutoff frequency for the post detection bandpass filter, resulting in more broadband noise for the same data rate. Thus, the decision depends on the assumptions of the type of interference to be encountered and even then the answer is not clear cut.

An analysis of the effect of different types of interference is given by Anthes of RF Monolithics (see references). He concludes that ASK, which does not completely shut off the carrier on a "0" bit, is marginally better than FSK.

Continuous digital communication

For efficient transmission of continuous digital data, the modulation choices are much more varied than in the case of event transmission. We can see this in the three leading cellular digital radio systems, all of which have the same use and basic requirements. The system referred to as D-AMPS or TDMA (time division multiple access) uses a type of modulation called Pi/4 DPSK (differential phase shift keying). The GSM network is based on GMSK (Gaussian minimum shift keying). The third major system is CDMA (code division multiple access). Each system claims that its choice is best, but it is clear that there is no simple cut-and-dried answer. We aren't going into the details of the cellular systems here, so we'll look at the relevant trade-offs for modulation methods in what we have defined as short-range radio applications. At the end of this chapter, we review the basic principles of digital modulation and spread-spectrum modulation.

For the most part, license-free applications specify ISM bands where signals are not confined to narrow bandwidth channels. However, noise power is directly proportional to bandwidth, so the receiver bandwidth should be no more than is required for the data rate used. Given a data rate and an average or peak power limitation, there are several reasons for preferring one type of modulation over another. They involve error rate, implementation complexity, and cost. A common way to compare performance of the different systems is by curves of bit error rate (BER) versus the signal-to-noise ratio, expressed as energy per bit divided by the noise density (defined below). The three most common types of modulation system are compared in Figure 4-9. Two of the modulation types were mentioned above. The third, phase shift keying, is described below.

Phase shift keying (PSK)

Whereas in amplitude shift keying and frequency shift keying the amplitude and frequency are varied according to the digital source data, in PSK it is the phase of the RF carrier that is varied. In its simplest form, the waveform looks like Figure 4-10. Note that the phase of the carrier

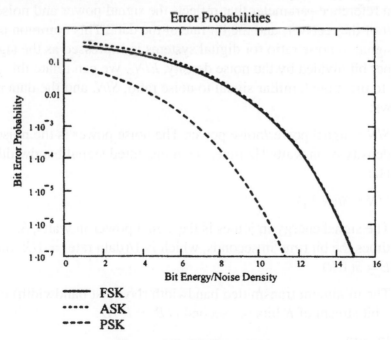

Figure 4-9: Bit Error Rates

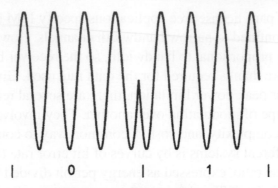

| | 0 | | 1 | |

Figure 4-10: Phase Shift Keying

wave shifts 180 deg. according to the data signal bits. Similar to FSK, the carrier remains constant, thus giving the same advantage that we mentioned for FSK—maximum signal-to-noise ratio when there is a peak power limitation.

Comparing digital modulation methods

In comparing the different digital modulation methods, we need a common reference parameter that reflects the signal power and noise at the input of the receiver, and the bit rate of the data. This common parameter of signal to noise ratio for digital systems is expressed as the signal energy per bit divided by the noise density, E/N_o. We can relate this parameter to the more familiar signal to noise ratio, S/N, and the data rate R, as follows:

1) S/N = signal power/noise power. The noise power is the noise density N_o in watts/Hz times the transmitted signal bandwidth B_T in Hz:

   $S/N = S/(N_oB_T)$

2) The signal energy in joules is the signal power in watts, S, times the bit time in seconds, which is 1/(data rate) = $1/R$, thus $E = S(1/R)$

3) The minimum transmitted bandwidth (Nyquist bandwidth) to pass a bit stream of R bits per second is $B_T = R$ Hz.

4) In sum:

   $E/N_o = (S/R)/N_o = S/N_oR = S/N$

The signal power for this expression is the power at the input of the receiver, which is derived from the radiated transmitted power, the receiver antenna gain, and the path loss. The noise density N_o, or more precisely the one sided noise power spectral density, in units of watts/Hz, can be calculated from the expression:

$$N_o = kT = 1.38 \text{ x } 10^{-23} \times T(\text{Kelvin}) \quad \text{watt/hertz}$$

The factor k is Boltzmann's constant and T is the equivalent noise temperature that relates the receiver input noise to the thermal noise that is present in a resistance at the same temperature T. Thus, at standard room temperature of 290 degrees Kelvin, the noise power density is 4×10^{-21} watts/Hz, or −174 dBm/Hz.

The modulation types making up the curves in Figure 4-9 are

Phase shift keying (PSK)

Noncoherent frequency shift keying (FSK)

Noncoherent Amplitude shift keying (ASK)

We see from the curves that the best type of modulation to use from the point of view of lowest bit error for a given signal-to-noise ratio (E/N_o) is PSK. There is essentially no difference, according to the curves, between ASK and FSK. (This is true only when noncoherent demodulation is used, as in most simple short range systems). What then must we consider in making our choice?

PSK is not difficult to generate. It can be done by a balanced modulator. The difficulty is in the receiver. A balanced modulator can be used here too but one of its inputs, which switches the polarity of the incoming signal, must be perfectly correlated with the received signal carrier and without its modulation. The balanced modulator acts as a multiplier and when a perfectly synchronized RF carrier is multiplied by the received signal, the output, after low-pass filtering, is the original bit stream. PSK demodulation is shown in Figure 4-11.

There are several ways of generating the required reference carrier signal from the received signal. Two examples are the Costas loop and the squaring loop, which include three multiplier blocks and a variable frequency oscillator (VFO). [See Dixon.] Because of the complexity and cost, PSK is not commonly used in inexpensive short-range equipment,

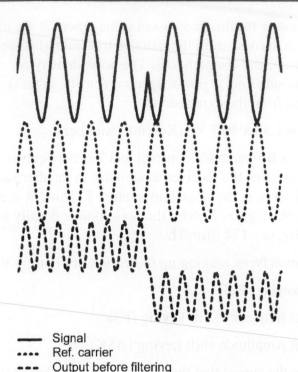

—— Signal
•••• Ref. carrier
--- Output before filtering

Figure 4-11: PSK Demodulation

but it is the most efficient type of modulation for high-performance data communication systems.

Amplitude shift keying is easy to generate and detect, and as we see from Figure 4-9, its bit error rate performance is essentially the same as for FSK. However, FSK is usually the modulation of choice for many systems. The primary reason is that peak power is usually the limitation, which gives FSK a 3-dB advantage, since it has constant power for both bit states, whereas ASK has only half the average power, assuming a 50% duty cycle and equal probability of marks and spaces. FSK has slightly more complexity than ASK, and that's probably why it isn't used in all short-range digital systems.

A modulation system which incorporates the methods discussed above but provides a high degree of interference immunity is spread spectrum, which we'll discuss later on in this chapter.

Analog communication

For short-range analog communication—wireless microphones, wireless earphones, auditive assistance devices—FM is almost exclusively used. When transmitting high-quality audio, FM gives an enhanced post-detection signal-to-noise ratio, at the expense of greater bandwidth. Even for narrow band FM, which doesn't have post-detection signal-to-noise enhancement, its noise performance is better than that of AM, because a limiting IF amplifier can be used to reduce the noise. AM, being a linear modulation process, requires linear amplifiers after modulation in the transmitter, which are less efficient than the class C amplifiers used for FM. Higher power conversion efficiency gives FM an advantage in battery-operated equipment.

Advanced digital modulation

Two leading characteristics of wireless communication in the last few years are the need for increasing data rates and better utilization of the radio spectrum. This translates to higher speeds on narrower bandwidths. At the same time, much of the radio equipment is portable and operated by batteries. So what is needed is:

- high data transmission rates
- narrow bandwidth
- low error rates at low signal-to-noise ratios
- low power consumption

Breakthroughs have occurred with the advancement of digital modulation and coding systems. We deal here with digital modulation principles. Coding will be discussed in Chapter 10.

The types of modulation that we discussed previously, ASK, FSK, and PSK, involve modifying a radio frequency carrier one bit at a time. We mentioned the Nyquist bandwidth, which is the narrowest bandwidth of an ideal filter which permits passing a bit stream without intersymbol interference. As the bandwidth of the digital bit stream is further reduced, the bits are lengthened and interfere with the detection of subsequent bits. This minimum, or Nyquist, bandwidth equals one-half of the bit rate at

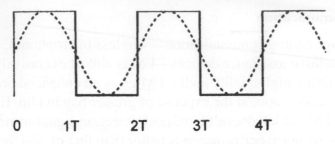

Figure 4-12: Nyquist Bandwidth

baseband, but twice as much for the modulated signal. We can see this result in Figure 4-12. An alternating series of marks and spaces can be represented by a sine wave whose frequency is one-half the bit rate: $f_{sin} = 1/2T$. An ideal filter with a lower cutoff frequency will not pass this fundamental frequency component and the data will not get through.

Any other combination of bits will create other frequencies, all of which are lower than the Nyquist frequency. It turns out then that the maximum number of bits per hertz of filter cutoff frequency that can be passed at baseband is two. Therefore, if the bandwidth of a telephone line is 3.4 kHz, the maximum binary bit rate that it can pass without intersymbol interference is 6.8k bits per second. We know that telephone line modems pass several times this rate. They do it by incorporating several bits in each *symbol* transmitted, for it is actually the symbol rate that is limited by the Nyquist bandwidth, and the problem that remains is to put several bits on each symbol in such a manner that they can be effectively taken off the symbol at the receiving end with as small as possible chance of error, given a particular S/N.

In Figure 4-13 we see three ways of combining bits with individual symbols, each of them based on one of the basic types of modulation—ASK, FSK, PSK. Each symbol duration T can carry one of four different values, or two bits. Using any one of the modulation types shown, the telephone line, or wireless link, can pass a bit rate twice as high as before over the same bandwidth. Combinations of these types are also employed, particularly of ASK and PSK, to put even more bits on a symbol. Quadrature amplitude modulation, QAM, sends several signal levels on four phases of the carrier frequency to give a high bandwidth efficiency—a high bit-rate relative to the signal bandwidth. It seems then that there is essentially no limit to the number of bits that could be compressed into a given bandwidth. If that were true, the 3.4-kHz telephone line could carry millions of bits per second, and the internet bottleneck to our homes would no longer exist. However, there is a very definite limit to the rate of information transfer over a transmission medium where noise is present, expressed in the Hartley-Shannon law:

$$C = W \log(1 + S/N)$$

This expression tells us that the maximum rate of information (the capacity C) that can be sent without errors on a communication link is a function of the bandwidth, W, and the signal-to-noise ratio, S/N. We examine this equation in greater detail in Chapter 10.

In investigating the ways of modulating and demodulating multiple bits per symbol, we'll first briefly discuss a method not commonly used in short-range applications (although it could be). This is called M-ary FSK (Figure 4-13b) and in contrast to the aim we mentioned above of increasing the number of bits per hertz, it increases the required bandwidth as the number of bits per symbol is increased. "M" in "M-FSK" is the number of different

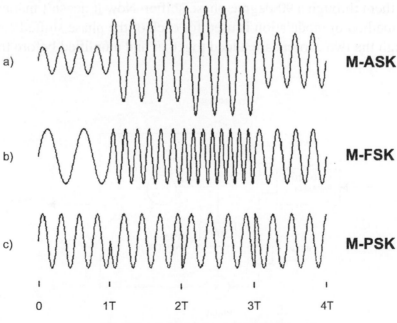

Figure 4-13: M-ary Modulation

frequencies that may be transmitted in each symbol period. The benefit of this method is that the required S/N per bit for a given bit error rate decreases as the number of bits per symbol increases. This is analogous to analog FM modulation, which uses a wideband radio channel, well in excess of the bandwidth of the source audio signal, to increase the resultant S/N. M-ary FSK is commonly used in point-to-point microwave transmission links for high-speed data communication where bandwidth limitation is no problem but the power limitation is.

Most of the high-data-rate bandwidth-limited channels use multiphase PSK or QAM. While there are various modulation schemes in use, the essentials of most of them can be described by the block diagram in Figure 4-14. This diagram is the basis of what may be called vector modulation, IQ modulation, or quadrature modulation. "I" stands for "in phase" and "Q" stands for "quadrature."

The basis for quadrature modulation is the fact that two completely independent data streams can be simultaneously modulated on the same frequency and carrier wave. This is possible if each data stream modulates coherent carriers whose phases are 90 degrees apart. We see in the diagram that each of these carriers is created from the same source by passing one of them through a 90-degree phase shifter. Now it doesn't matter which method of modulation is used for each of the phase shifted carriers. Although the two carriers are added together and amplified before trans-

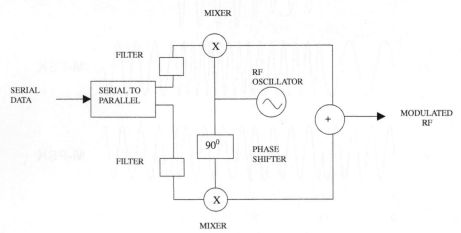

Figure 4-14: Quadrature Modulation

mission, the receiver, by reversing the process used in the transmitter (see Figure 4-15), can completely separate the incoming signal into its two components.

The diagrams in Figures 4-14 and 4-15 demonstrate quadrature phase shift keying (QPSK) where a data stream is split into two, modulated independently bit by bit, then combined to be transmitted on a single carrier. The receiver separates the two data streams, then demodulates and combines them to get the original serial digital data. The phase changes of the carrier in response to the modulation is commonly shown on a vector or constellation diagram. The constellation diagram for QPSK is shown in Figure 4-16. The Xs on the plot are tips of vectors that represent the magnitude (distance from the origin) and phase of the signal that is the sum of the "I" and the "Q" carriers shown on Figure 4-14, where each is multiplied by −1 or +1, corresponding to bit values of 0 and 1. The signal magnitudes of all four possible combinations of the two bits is the same— $\sqrt{2}$ when the I and Q carriers have a magnitude of 1. I and Q bit value combinations corresponding to each vector are shown on the plot.

We have shown that two data streams, derived from a data stream of rate R2, can be sent in parallel on the same RF channel and at the same bandwidth as a single data stream having half the bit rate, R1. At the receiver end, the demodulation process of each of the split bit streams is exactly the same as it would be for the binary phase shift modulation shown

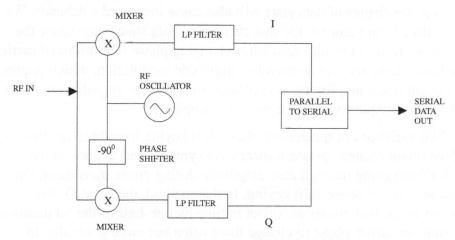

Figure 4-15: Quadrature Demodulation

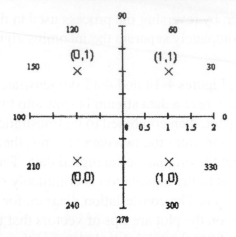

Figure 4-16: QPSK Constellation Diagram

in Figure 4-11 above, and its error rate performance is the same as is shown for the BPSK curve in Figure 4-9. Thus, we've doubled the data rate on the same bandwidth channel while maintaining the same error rate as before. In short, we've doubled the efficiency of the communication.

However, there are some complications in adopting quadrature modulation. Similar to the basic modulation methods previously discussed, the use of square waves to modulate RF carriers causes unwanted sidebands that may exceed the allowed channel bandwidth. Thus, special low-pass filters, shown in Figure 4-14, are inserted in the signal paths before modulation. Even with these filters, the abrupt change in the phase of the RF signal at the change of data state will also cause increased sidebands. We can realize from Figure 4-13c that changes of data states may cause the RF carrier to pass through zero when changing phase. Variations of carrier amplitude make the signal resemble amplitude modulation, which requires inefficient linear amplifiers, as compared to nonlinear amplifiers that can be used for frequency modulation, for example.

Two variations of quadrature phase shift keying have been devised to reduce phase changes between successive symbols and to prevent the carrier from going through zero amplitude during phase transitions. One of these is offset phase shift keying. In this method, the I and Q data streams in the transmitter are offset in time by one-half of the bit duration, causing the carrier phase to change more often but more gradually. In other words, the new "I" bit will modulate the cosine carrier at half a bit time earlier (or later) than the time that the "Q" bit modulates the sine carrier.

The other variant of QPSK is called pi/4 DPSK. It is used in the US TDMA (time division multiple access) digital cellular system. In it, the constellation diagram is rotated pi/4 radians (45 degrees) at every bit time such that the carrier phase angle can change by either 45 or 135 degrees. This system reduces variations of carrier amplitude so that more efficient nonlinear power amplifiers may be used in the transmitter.

Another problem with quadrature modulation as described above is the need for a coherent local oscillator in the receiver in order to separate the in-phase and quadrature data streams. As for bipolar phase shift keying, this problem may be ameliorated by using differential modulation and by multiplying a delayed replica of the received signal by itself to extract the phase differences from symbol to symbol.

The principle of transmitting separate data streams on in phase and quadrature RF carriers may be extended so that each symbol on each carrier contains 2, 3, 4 or more bits. For example, 4 bits per each carrier vector symbol allows up to 16 amplitude levels per carrier and a total of 256 different states of amplitude and phase altogether. This type of modulation is called quadrature amplitude modulation (QAM), and in this example, 256-QAM, 8 data bits can be transmitted in essentially the same time and bandwidth as one bit sent by binary phase shift keying. The constellation of 16-QAM (4 bits per symbol) is shown in Figure 4-17.

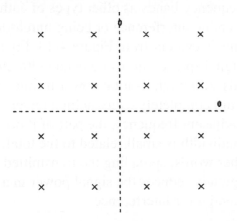

Figure 4-17: 16-QAM Constellation Diagram

Remember that the process of concentrating more and more bits to a carrier symbol cannot go on without limit. While the bit rate per hertz goes up, the required S/N of the channel for a given bit error rate (BER) also increases—that is, more power must be transmitted, or as a common alternative in modern digital communication systems, sophisticated coding algorithms are incorporated in the data protocol. However, the ultimate limit of data rate for a particular communication channel with noise is the Hartley-Shannon limit stated above.

We now turn to examine another form of digital modulation that is becoming very important in a growing number of short-range wireless applications—spread-spectrum modulation.

Spread spectrum

The regulations for unlicensed communication using unspecified modulation schemes, both in the US and in Europe, determine maximum power outputs ranging from tens of microwatts up to 10 milliwatts in most countries. This limitation greatly reduces the possibilities for wireless devices to replace wires and to obtain equivalent communication reliability. However, the availability of frequency bands where up to one watt may be transmitted in the US and 100 mW in Europe, under the condition of using a specified modulation system, greatly enlarges the possible scope of use and reliability of unlicensed short-range communication.

Spread spectrum has allowed the telecommunication authorities to permit higher transmitter powers because spread-spectrum signals can coexist on the same frequency bands as other types of authorized transmissions without causing undue interference or being unreasonably interfered with. The reason for this is evident from Figure 4-18. Figure 4-18a shows the spread-spectrum signal spread out over a bandwidth much larger than the narrow-band signals. Although its total power if transmitted as a narrow band signal could completely overwhelm another narrow-band signal on the same or adjacent frequency, the part of it occupying the narrow-band signal bandwidth is small related to the total, so it doesn't interfere with it. In other words, spreading the transmitted power over a wide frequency band greatly reduces the signal power in a narrow bandwidth and thus the potential for interference.

Figure 4-18b shows how spread-spectrum processing reduces interference from adjacent signals. The despreading process concentrates the total power of the spread-spectrum signal into a narrow-band high peak power signal, whereas the potentially interfering narrow-band signals are spread out so that their power in the bandwidth of the desired signal is relatively low.

These are some advantages of spread spectrum modulation:

- FCC rules allow higher power for nonlicensed devices
- Reduces co-channel interference—good for congested ISM bands
- Reduces multipath interference
- Resists intentional and unintentional jamming

a) Signals at receiver input

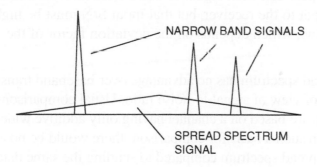

b) Signals after despreading

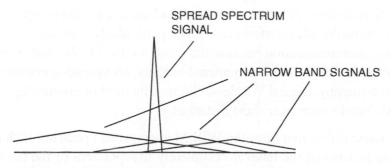

Figure 4-18: Spread Spectrum and Narrow-Band Signals

- Reduces the potential for eavesdropping

- Permits code division multiplexing of multiple users on a common channel.

There is sometimes a tendency to compare spread spectrum with wide-band frequency modulation, such as used in broadcasting, since both processes achieve performance advantages by occupying a channel bandwidth much larger than the bandwidth of the information being transmitted. However, it's important to understand that there are principal differences in the two systems.

In wide band FM (WBFM), the bandwidth is spread out directly by the amplitude of the modulating signal and at a rate determined by its frequency content. The result achieved is a signal-to-noise ratio that is higher than that obtainable by sending the same signal over baseband (without modulation) and with the same noise density as on the RF channel. In WBFM, the post detection signal-to-noise ratio (S/N) is a multiple of the S/N at the input to the receiver, but that input S/N must be higher than a threshold value, which depends on the deviation factor of the modulation.

In contrast, spread spectrum has no advantage over baseband transmission from the point of view of signal-to-noise ratio. Usual comparisons of modulation methods are based on a channel having only additive wideband Gaussian noise. With such a basis for comparison, there would be no advantage at all in using spread spectrum compared to sending the same data over a narrow-band link. The advantages of spread spectrum are related to its relative immunity to interfering signals, to the difficulty of message interception by a chance eavesdropper, and to its ability to use code selective signal differentiation. Another often-stated advantage to spread spectrum—reduction of multipath interference—is not particularly relevant to short-range communication because the pulse widths involved are much longer than the delay times encountered indoors. (A spread-spectrum specialist company, Digital Wireless, claims a method of countering multipath interference over short distances.)

The basic difference between WBFM and spread spectrum is that the spreading process of the latter is completely independent of the baseband signal itself. The transmitted signal is spread by one (or more) of several different spreading methods, and then un-spread in a receiver that knows the spreading code of the transmitter.

The methods for spreading the bandwidth of the spread-spectrum transmission are frequency-hopping spread spectrum (FHSS), direct-sequence spread spectrum (DSSS), pulsed-frequency modulation or chirp modulation, and time-hopping spread spectrum. The last two types are not allowed in the FCC rules for unlicensed operation and after giving a brief definition of them we will not consider them further.

(1) In frequency-hopping spread spectrum, the RF carrier frequency is changed relatively rapidly at a rate of the same order of magnitude as the bandwidth of the source information (analog or digital), but not dependent on it in any way. At least several tens of different frequencies are used, and they are changed according to a pseudo-random pattern known also at the receiver. The spectrum bandwidth is roughly the number of the different carrier frequencies times the bandwidth occupied by the modulation information on one hopping frequency.

(2) The direct-sequence spread-spectrum signal is modulated by a pseudo-random digital code sequence known to the receiver. The bit rate of this code is much higher than the bit rate of the information data, so the bandwidth of the RF signal is consequently higher than the bandwidth of the data.

(3) In chirp modulation, the transmitted frequency is swept for a given duration from one value to another. The receiver knows the starting frequency and duration so it can unspread the signal.

(4) A time-hopping spread-spectrum transmitter sends low duty cycle pulses with pseudo-random intervals between them. The receiver unspreads the signal by gating its reception path according to the same random code as used in the transmitter.

Actually, all of the above methods, and their combinations that are sometimes employed, are similar in that a pseudo-random or arbitrary (in the case of chirp) modulation process used in the transmitter is duplicated in reverse in the receiver to unravel the wide-band transmission and bring it to a form where the desired signal can be demodulated like any narrow-band transmission.

The performance of all types of spread-spectrum signals is strongly related to a property called process gain. It is this process gain that quantifies the degree of selection of the desired signal over interfering narrow-band and other wide-band signals in the same passband. Process gain is the difference in dB between the output S/N after unspreading and the input S/N to the receiver:

$$PG_{dB} = (S/N)_{out} - (S/N)_{in}$$

The process gain factor may be approximated by the ratio

$$PG_f = \text{(RF bandwidth)/(rate of information)}$$

A possibly more useful indication of the effectiveness of a spread-spectrum system is the jamming margin:

$$\text{Jamming Margin} = PG - (L_{sys} + (S/N)_{out})$$

where L_{sys} is system implementation losses, which may be of the order of 2 dB.

The jamming margin is the amount by which a potentially interfering signal in the receiver's passband may be stronger than the desired signal without impairing the desired signal's ability to get through.

Let's now look at the details of frequency hopping and direct-sequence spread spectrum.

Frequency hopping

FHSS can be divided into two classes—fast hopping and slow hopping. A fast-hopping transmission changes frequency one or more times per data bit. In slow hopping, several bits are sent per hopping frequency. Slow hopping is used for fast data rates since the frequency synthesizers in the transmitter and receiver are not able to switch and settle to new frequencies fast enough to keep up with the data rate if one or fewer bits per hop are transmitted. The spectrum of a frequency-hopping signal looks like Figure 4-19.

Figure 4-20 is a block diagram of FHSS transmitter and receiver. Both transmitter and receiver local oscillator frequencies are controlled by frequency synthesizers. The receiver must detect the beginning of a transmission and synchronize its synthesizer to that of the transmitter. When the receiver knows the pseudo-random pattern of the transmitter, it can lock onto the incoming signal and must then remain in synchronization by changing frequencies at the same time as the transmitter. Once exact synchronization has been obtained, the IF frequency will be constant and the signal can be demodulated just as in a normal narrow-band superheterodyne receiver. If one or more of the frequencies that the transmission occupies momentarily is also occupied by an interfering signal, the bit or bits that were transmitted at that time may be lost. Thus, the transmitted message must contain redundancy or error correction coding so that the lost bits can be reconstructed.

Direct Sequence

Figure 4-21 is a diagram of a DSSS system. A pseudo-random spreading code modulates the transmitter carrier frequency, which is then modulated by the data. The elements of this code are called chips. The frequency spectrum created is shown in Figure 4-22. The required bandwidth is the width of the major lobe, shown on the drawing as $2 \times Rc$, or twice the chip rate. Due to the wide bandwidth, the signal-to-noise ratio at the receiver input is very low, often below 0 dB. The receiver multiplies a replica of the transmitter pseudo-random spreading code with the receiver local oscillator and the result is mixed with the incoming signal. When the

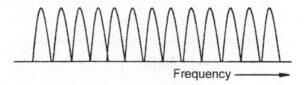

Figure 4-19: Spectrum of Frequency-Hopping Spread Spectrum Signal

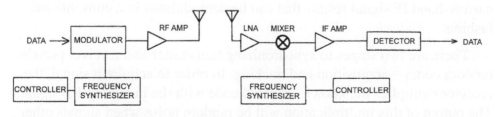

Figure 4-20: FHSS Transmitter and Receiver

113

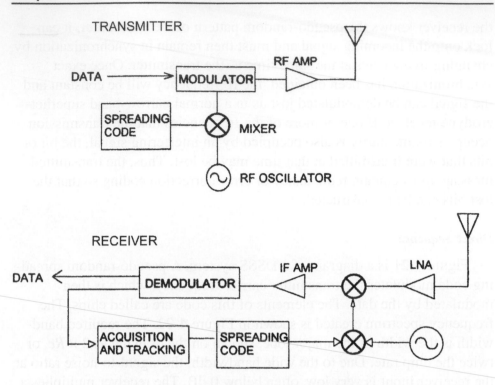

TRANSMITTER

DATA → MODULATOR

RF AMP

SPREADING CODE

MIXER

RF OSCILLATOR

RECEIVER

DATA ← DEMODULATOR

IF AMP

LNA

ACQUISITION AND TRACKING

SPREADING CODE

Figure 4-21: DSSS Transmitter and Receiver

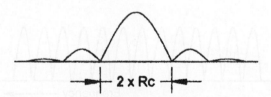

$2 \times Rc$

Figure 4-22: DSSS Frequency Spectrum

transmitter and receiver spreading codes are the same and are in phase, a narrow-band IF signal results that can be demodulated in a conventional fashion.

There are two stages to synchronizing transmitter and receiver pseudo-random codes—acquisition and tracking. In order to acquire a signal, the receiver multiplies the known expected code with the incoming RF signal. The output of this multiplication will be random noise when signals other than the desired signal exist on the channel. When the desired signal is present, it must be synchronized precisely in phase with the receiver's code sequence to achieve a strong output. A common way to obtain this synchronization is for the receiver to adjust its code to a rate slightly different from that of the transmitter. Then the phase difference between the two codes will change steadily until they are within one bit of each other, at which time the output of the multiplier increases to a peak when the codes are perfectly synchronized. This method of acquisition is called a sliding correlator.

The next stage in synchronization, tracking, keeps the transmitted and received code bits aligned for the complete duration of the message. In the method called "tau delta" the receiver code rate is varied slightly around the transmitted rate to produce an error signal that is used to create an average rate exactly equal to the transmitter's.

Other methods for code acquisition and tracking can be found in any book on spread spectrum—for example, Dixon's—listed in the references. Once synchronization has been achieved, the resulting narrow-band IF signal has a S/N equal to the received S/N plus the process gain, and it can be demodulated as any narrow-band signal. The difference in using spread spectrum is that interfering signals that would render the normal narrow-band signal unintelligible are now reduced by the amount of the process gain.

Relative advantages of DSSS and FHSS

The advantages of one method over the other are often debated, but there is no universally agreed-upon conclusion as to which is better, and both types are commonly used. They both are referenced in the IEEE specification 802.11 for wireless LANs. However some conclusions, possibly debatable, can be made:

- For a given data rate and output power, an FHSS system can be designed with less power consumption than DSSS.

- DSSS can provide higher data rates with less redundancy.

- Acquisition time is generally lower with FHSS (although short times are achieved with DSSS using matched filter and short code sequence).

- FHSS may have higher interference immunity, particularly when compared with DSSS with lowest allowed process gain.

- DSSS has more flexibility—frequency agility may be used to increase interference immunity.

4.5 RFID

A growing class of applications for short-range wireless is radio frequency identification (RFID). A basic difference between RFID and the applications discussed above is that RFID devices are not communication devices *per se* but involve interrogated transponders. Instead of having two separate transmitter and receiver terminals, an RFID system consists of a reader that sends a signal to a passive or active tag and then receives and interprets a modified signal reflected or retransmitted back to it. See Figure 1-3.

The reader has a conventional transmitter and receiver having similar characteristics to the devices already discussed. The tag itself is special for this class of applications. It may be passive, receiving its power for re-transmission from the signal sent from the reader, or it may have an active receiver and transmitter and tiny embedded long-life battery. Ranges of RFID may be several centimeters up to tens of meter. Operation frequencies range from 125 kHz up to 2.45 GHz. Frequencies are usually those specified for nonlicensed applications. Higher data rates demand higher frequencies.

In its most common form of operation, an RFID system works as follows. The reader transmits an interrogation signal, which is received by the tag. The tag may alter the incoming signal in some unique manner and reflect it back, or its transmitting circuit may be triggered to read its ID code residing in memory and to transmit this code back to the receiver. Active tags have greater range than passive tags. If the tag is moving in relation to the receiver, such as in the case of toll collection on a highway, the data transfer must take place fast enough to be complete before the tag is out of range.

These are some design issues for RFID:

- Tag orientation is likely to be random, so tag and reader antennas must be designed to give the required range for any orientation.

- Multiple tags within the transmission range can cause return message collisions. One way to avoid this is by giving the tags random delay times for response, which may allow several tags to be interrogated at once.

- Tags must be elaborately coded to prevent misreading.

4.6 Summary

This chapter has examined various characteristics of short-range systems. It started with the ways in which data is formatted into different information fields for transmission over a wireless link. We looked at several methods of encoding the one's and zero's of the baseband information before modulation in order to meet certain performance requirements in the receiver, such as constant DC level and minimum bit error probability. Analog systems were also mentioned, and we saw that pre-emphasis/de-emphasis and compression/expansion circuits in voice communication devices improve the signal-to-noise ratio and increase the dynamic range.

Reasons for preferring frequency or amplitude digital modulation were presented from the points of view of equipment complexity and of the different regulatory requirements in the US and in Europe. Similarly, there are several considerations in choosing a frequency band for a wireless system, among them background noise, antenna size, and cost.

The three basic modulation types involve impressing the baseband data on the amplitude, frequency, or phase of an RF carrier signal. Modern digital communication uses combinations and variations in the basic methods to achieve high bandwidth efficiency, or conversely, high signal-to-noise ratio with relatively low power. We gave an introduction to quadrature modulation and to the principles of spread-spectrum communication. The importance of advanced modulation methods is on the rise, and they can surely be expected to have an increasing influence on short-range radio design in the near future.

Transmitters

In this chapter we examine the details of transmitter design. The basic constituents of a radio transmitter are shown in Figure 5-1. Source data, which may be analog or digital, is imposed on a radio-frequency carrier wave by a modulator, then amplified, filtered, and applied to an antenna. The transmitter must also have a source of power. First we'll look at the definition of each block, and then see various ways in which the whole transmitter can be implemented.

5.1 RF Source

Short-range radio transmitters have one of four types of frequency control: *LC*, SAW, crystal, and synthesizer. Another type of control, direct digital synthesis, may show up in short-range devices in the future and so we will say some words about it.

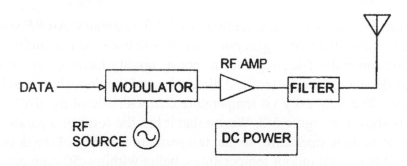

Figure 5-1: Basic Transmitter Block Diagram

In comparing the attributes of the several frequency-controlling devices, we refer to their accuracy and stability. Accuracy is the degree of deviation from a nominal frequency at a given temperature and particular circuit characteristics. Stability expresses how much the frequency may change over a temperature range and under other external conditions such as proximity to surrounding objects.

LC Control

Only the simplest and least expensive portable transmitters may be expected to employ an inductor-capacitor (*LC*) resonator frequency control source. It has largely been supplanted by the SAW resonator. Like the SAW device, the *LC* circuit directly generates the RF carrier, almost always in the UHF range. Its stability is in the order of hundreds of kHz, and it suffers from hand effect and other proximity disturbances, which can move the frequency. Because of its poor stability and accuracy, it is almost always used with a superregenerative receiver, which has a broad bandwidth response. The *LC* oscillator circuit is used in garage door openers and automobile wireless keys.

SAW Resonators

SAW stands for *surface acoustic wave*. Most short-range wireless control and alarm devices use it as the frequency determining element. It's more stable than the *LC* resonator, and simple circuits based on it have a stability of around ±30 kHz. Its popularity is based on its reasonable price and the fact that it generates relatively stable UHF frequencies directly from a simple circuit.

Figure 5-2 shows the construction of a SAW resonator. An *RF* voltage applied to metal electrodes generates an acoustic wave on the surface of the quartz substrate. The frequency response, quality factor Q, and insertion loss depend on the pattern and dimensions of the deposited metal electrodes. The frequency vs. temperature characteristic of the SAW device is shown in Figure 5-3. We see that it has the form of a parabola peaking at room temperature, so that temperature stability of the device is very good at normal indoor temperatures, being within – 50 ppm or 0.005% between 0 and 75 °C.

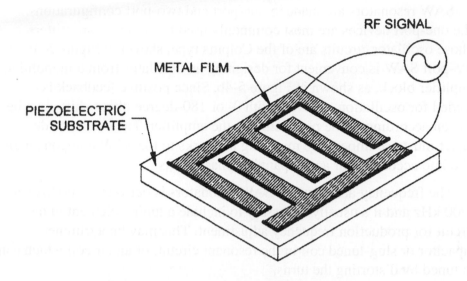

Figure 5-2: SAW Resonator Construction

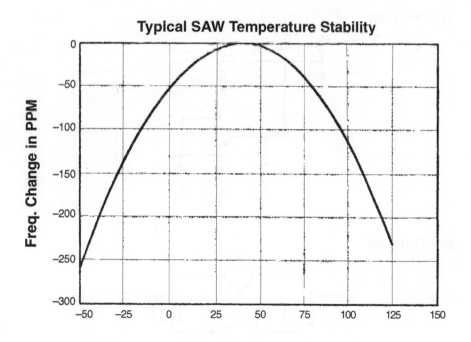

Figure 5-3: SAW Frequency vs. Temperature

121

SAW resonators are made in one-port and two-port configurations. The one-port devices are most commonly used for simple transmitters, whose oscillator circuits are of the Colpitts type, shown in Figure 5-4a. A two-port SAW is convenient for designing an oscillator from a monolithic amplifier block, as shown in Figure 5-4b. Since positive feedback is needed for oscillation, a device with 0- or 180-degree phase shift must be chosen according to the phase shift of the amplifier. The oscillator designer must use impedance matching networks to the SAW component and take into account their accompanying phase shifts.

The frequency accuracy for SAW resonators is between ±100 kHz and ±200 kHz and it's usually necessary to include a tuning element in the circuit for production frequency adjustment. This may be a trimmer capacitor or slug-tuned coil in the resonant circuit, or an air coil which can be tuned by distorting the turns.

In addition to resonators, SAW UHF filters are also available for short-range transmitters and receivers.

a) 1-Port SAW

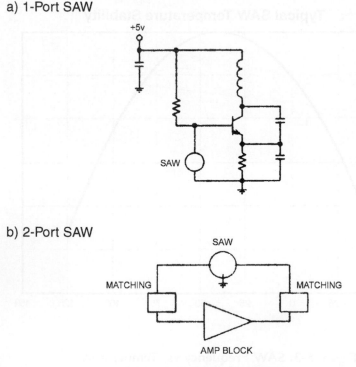

b) 2-Port SAW

Figure 5-4: SAW Oscillators

SAW components, both resonators and filters, are available from around 200 to 950 MHz. They cover frequencies used for unlicensed devices both in the US and Canada, the United Kingdom and continental Europe and selected other countries. You can obtain resonators at offset frequencies for local oscillators in superheterodyne receivers.

When designing a short-range radio system based on a SAW device, you should check the frequencies available from the various manufacturers for transmitter and receiver oscillators and filters before deciding on the frequencies you will use. Also take into account the countries where the product will be sold and the frequencies allowed there. Deviating from standard off-the-shelf frequencies should be considered only for very large production runs, since it entails special set-up charges and stocking considerations.

SAW resonators are available in metal cans and in SMT packages for surface mount assembly. There is no consistency in packages and terminal connections between manufacturers, a point that should be considered during design.

Crystal Oscillators

While the availability of SAW devices for short-range unlicensed applications resulted in a significant upgrading of performance compared to *LC* circuits, they still are not suitable for truly high performance uses. Sensitive receivers need narrow bandwidth, and this requires an accurate, stable local oscillator and correspondingly accurate transmitter oscillator. Stability in the range of several kHz is obtainable only with quartz crystals.

Crystals, like SAWs, are made of quartz piezoelectric material. In contrast to SAWs, whose operation we have seen is based on surface wave propagation, the crystal vibration is a bulk effect caused by high-frequency excitation to electrodes positioned on opposite sides of the bulk material. Crystal oscillators used for short-range transmitters are of two basic forms: fundamental mode and overtone mode. The fundamental mode usually extends from tens of kHz up to around 25 MHz, whereas overtone oscillators may operate up to 200 MHz. The overtone frequency is usually about three, five or seven times the fundamental frequency of the crystal.

In order to generate UHF frequencies in the range of 300 to 950 MHz, we must incorporate one or more doubler or tripler stages in the frequency-generating stage of the transmitter, or the local oscillator of the receiver. This involves additional components and space and explains why the SAW-based circuits are so much more popular for low-cost, short-range equipment, particularly hand-held and keychain transmitters.

However, considering the frequency multiplier stages of a crystal oscillator as a disadvantage compared to the SAW-based oscillator isn't always justified. The single-stage SAW oscillator, when coupled directly to an antenna, is subject to frequency pulling due to the proximity effect or hand effect. Conducting surfaces, including the human body, within centimeters of the antenna and resonant circuits can displace the frequency of a SAW oscillator by 20 kHz or more. To counter this effect, a buffer circuit is often needed to isolate the antenna from the oscillator. With this buffer, there is hardly a cost or board space difference between the SAW transmitter and crystal transmitter with one doubler or tripler stage.

Another factor in favor of the crystal oscillator transmitter is that it can be designed more easily without need for production tuning. Production frequency accuracy without tuning of up to 10 kHz is acceptable for many short-range applications.

A circuit for a crystal oscillator transmitter is shown in Figure 5-5.

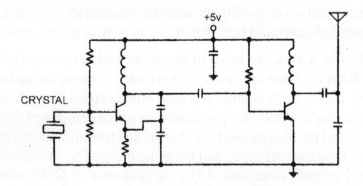

Figure 5-5: Crystal Oscillator Transmitter

Synthesizer Control

The basic role of a frequency synthesizer is to generate more than one frequency using only one frequency-determining element, usually a crystal. Other reasons for using a synthesizer are

- The output frequency is derived directly from an oscillator without using multipliers;

- It allows higher deviation in frequency modulation than when directly modulating a crystal oscillator.

A block diagram of a frequency synthesizer, also called a phase-locked loop (PLL), is shown in Figure 5-6. Its basic components are the reference crystal oscillator, a phase detector (PD), a low-pass filter, a voltage (or current) controlled oscillator (VCO), and a frequency divider. The reference divider and prescaler are added for reasons mentioned below.

The output frequency is generated by an oscillator operating on that frequency. Its exact frequency is controllable by a voltage or current input at the control terminal. The range of the control signal must be able to sweep the output over the frequency range desired for the device. A common way to vary the frequency of the VCO is to use a varactor diode, which has a capacitance vs. voltage transfer characteristic, in the frequency-determining resonant circuit of the oscillator.

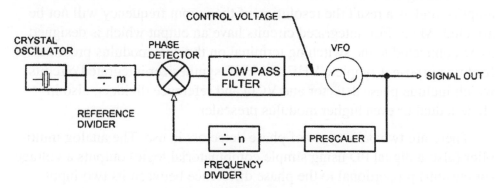

Figure 5-6: Frequency Synthesizer

The frequency synthesizer is a feedback control device, similar in many ways to a servo mechanism. An error signal is created in the phase detector, which outputs a voltage (or current) whose value is proportional to the difference in phase between a signal derived from the output signal, and the reference signal. Since the crystal reference frequency is much lower than the output frequency we want to stabilize, a frequency divider element is inserted between the VCO and the PD. The low-pass filter following the PD determines the stability of the PLL, its speed of response, and the rejection of spurious output frequencies and noise.

The division ratio, n, of the frequency divider determines the output frequency, since the control voltage will adjust itself through the feedback such that f_{out}/n will equal the crystal reference frequency. Increasing or decreasing n by 1 increases or decreases the output frequency by the amount of the reference frequency. The reference divider is inserted in the circuit when frequency resolution smaller than the value of the crystal is needed. The prescaler, which is a low value frequency divider, is required when the maximum input frequency specified for the variable ratio divider is lower than the output frequency. The use of a prescaler having a single frequency division ratio increases the incremental frequency of the synthesizer by a factor equal to the prescaler division ratio. In order not to lose the frequency resolution desired, or to be forced to divide the reference frequency further because of the prescaler, a dual modulus prescaler can be used. This prescaler allows switching between two adjacent division values during the countdown of the divider—32/33 or 64/65, for example—and as a result the resolution of the output frequency will not be affected. Many PLL integrated circuits have an output which is designed to be connected to the switching terminal on the dual modulus prescaler for this purpose. Complete VHF and UHF synthesizer integrated circuits, which include phase detector and VCO and reference dividers, also include a dual or even higher modulus prescaler.

There are two basic types of phase detectors in use. The analog multiplier (also a digital PD using simple combinatorial logic) outputs a voltage (or current) proportional to the phase difference between its two input signals. When these signals are not on the same frequency, the phase difference will vary at a rate equal to the difference of the two frequencies. The negative feedback of the loop will act to lock the VCO frequency such that its divided frequency will equal the reference frequency and the phase difference will create exactly the voltage needed on the VCO control input to give the desired frequency. The phase difference at the PD during lock is equal to or very near 90 degrees.

Because of the low-pass filter, if the reference and VCO divided frequencies are too far apart, the rapidly varying phase signal can't get through to the VCO control and the synthesizer will not lock. The difference between the minimum and the maximum frequencies of the range over which the output frequency can be brought into lock is called the capture range. The output frequency remains locked as long as the open loop frequency remains within what is defined as the lock range. The lock range is greater than the capture range, and affects the degree of immunity of the frequency synthesizer to influences of changing temperature and other environmental and component variations that determine the stability of an oscillator. When the analog type phase detector is used, an auxiliary frequency discriminator circuit may be required to bring the VCO frequency into the range where lock can be obtained.

The other type of phase detector is what is called a charge-pump PLL. It's a digital device made from sequential logic circuitry (flip-flops). It works by sending voltage (or current) pulses to the low-pass filter, which smooths the pulses before application to the VCO. The polarity of these pulses and their width depend on the relationship between the digital signal edge transitions of the reference oscillator and the divided down VCO signals at its input. In contrast to the analog phase detector, the charge-pump PLL always locks both of its input signals to zero phase difference between them. This device acts as both frequency discriminator and phase detector so an auxiliary frequency discriminator is not required.

The characteristics of the low-pass filter determine the transient response of the synthesizer, the capture range, capture time, and rejection of noise originating in the VCO and the reference frequency source. Its design requires some compromises according to system requirements. For example, a relatively low cut-off frequency reduces spurious outputs due to modulation of the carrier by the reference source, and allows lower baseband frequencies with frequency modulation applied to the VFO.

However, the average time to obtain lock is relatively long, and the output signal passes more noise originating in the VCO and its control circuit. A higher cut-off frequency allows the loop to lock in quicker but reference oscillator noise and spurious frequencies will be more prominent in the output. As mentioned, filter design affects the use of frequency modulation, specifically analog. Changing of the output frequency is exactly what the PLL is supposed to prevent, so to allow the modulation to get through, the spectrum of the modulating signal, applied to the frequency control input of the VCO, must be above the cut-off frequency of the low-pass filter.

Although the output signal from a frequency synthesizer comes directly from an oscillator operating on the output frequency, it does contain spurious frequencies attributable to the PLL and in particular to the reference oscillator and its harmonics. A clean, noise-free signal is particularly important for the local oscillator of a superheterodyne receiver, since the spurious signals on the oscillator output can mix with undesired signals in the receiver passband and cause spurious responses and interference. Careful design and layout of the PLL is necessary to reduce them.

A reduction of spurious frequencies, or "spurs" as they are often called, on the synthesizer output is attainable using a fractional division or fractional-*N* synthesizer. This design also achieves fast switching, so it's used in frequency-hopping spread spectrum transmitters and receivers. The fractional-*N* synthesizer uses a complex manipulation of the dual modulus prescaler to create a fractional division ratio *N*. This technique permits switching the output frequency in increments that are smaller than the reference frequency. It is the higher reference frequency, compared to a normal synthesizer with the same frequency increments, that gives the fractional-*N* synthesizer its improved performance.

Direct Digital Synthesis

A method for radio frequency wave generation that allows precise control over frequency, phase and amplitude is direct digital synthesis. The direct digital synthesizer is not common in short-range applications but it is likely to be used in the future as devices become more sophisticated and DDS component prices come down. DDS involves generating waveforms digitally instead of being based on an analog oscillator as in the other methods described above. Digital words representing samples of a sine wave are applied to a digital-to-analog converter, whose output is low-pass filtered.

Output frequencies of DDS systems are in the range of units to tens of MHz, so they don't directly generate the frequencies required for most short-range devices. However, when used as the reference frequency in PLLs, UHF and microwave signals can be established with high accuracy and resolution. Amplitude, frequency and phase modulation can be applied directly by software to the generated signals, which gives great flexibility where complicated and precise waveforms are required.

5.2 Modulation

In this section we look at the most common methods of implementing ASK, FSK, PSK and FM in short-range transmitters.

ASK

Amplitude shift keying can be applied to *LC* and SAW oscillators by keying the oscillator stage bias, as shown in Figure 5-7. Resistor Rb is chosen to provide the proper bias current from the code generator, and by varying its value the power from the stage can be regulated in order to meet the requirements of the applicable specifications.

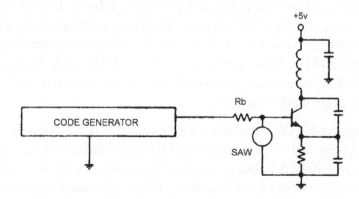

Figure 5-7: Amplitutde Shift Keying of SAW Oscillator

Switching the oscillator on and off works satisfactorily for low rates of modulation—up to around 2 kbits/second. At higher rates, the delay time needed for the oscillations to turn on will distort the transmitted signal and may increase the error rate at the receiver. In this case a buffer or amplifier stage following the oscillator must be keyed while the oscillator stage runs continuously. The problem in this situation with low-power transmitters is that the continuous signal from the oscillator may leak through the buffer stage when it is turned off, or may be radiated directly, reducing the difference between the ON and OFF periods at the receiver. To reduce this effect, the oscillator layout must be as compact as possible and its coil in particular must be small. In many cases the oscillator circuit will have to be shielded.

FSK

Frequency shift keying of a SAW oscillator can be achieved by switching a capacitance in and out of the tuned circuit in step with the modulation. A simple and inexpensive way to do it is shown in Figure 5-8. You can also use a transistor instead of a diode to do the switching, or use a varicap diode, whose capacitance changes with the applied voltage. The effect of changing the capacitance in the tuned circuit is to pull the oscillation frequency away from its static value, which is held by the SAW resonator. Deviation of several thousand hertz can be obtained by this circuit but the actual value depends on the insertion loss of the individual resonator. As we have seen in Section 5.1, SAW-controlled devices can vary as much as 30 kHz due to component aging and proximity effects. This forces compromises in the design of the receiver frequency discriminator, which must accept variable deviation of say 5 to 10 kHz while the center frequency may change 30 kHz. Thus, the performance of FSK SAW devices is not likely to realize the advantages over ASK that are expected, and indeed FSK is not often used with SAWs.

An improved SAW oscillator FSK developed by RF Monolithics is shown in Figure 5-9. This circuit is not based on indefinite pulling, as is the design in Figure 5-8, but on varying the resonance of the SAW as seen by the circuit, because of the insertion of a small-valued capacitance in series with the SAW. In the RFM circuit, this capacitance is that of the reverse connected transistor (emitter and collector terminal are reversed)

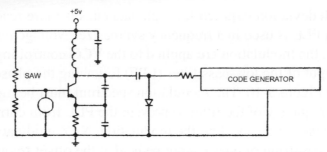

Figure 5-8: Frequency Shift Keying of SAW Oscillator

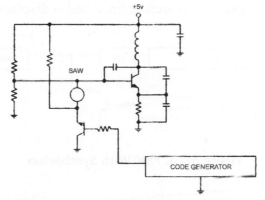

Figure 5-9: Improved FSK Modulator

when it is biased off. Reversing the transistor connection decreases its storage time, allowing a higher switching rate. RFM states that the circuit is patented and that licensing is required for using it (Reference 1).

If FSK is used with a SAW resonator, the oscillator must be buffered from the antenna and the oscillator should be shielded or otherwise designed to reduce the proximity effect to the minimum possible.

A crystal oscillator's frequency can be similarly pulled by detuning its tuned circuit or by switching a capacitor or inductor across or in series with the crystal terminals. A frequency deviation of several hundred hertz is possible. The frequency multiplier stages needed to step up the frequency to UHF also multiply the deviation. The required frequency deviation for effective demodulation is at least half the bit rate, so higher deviation and consequently more multiplication stages are needed for high data rates.

The high deviation required for high data rates is more readily obtainable when a PLL is used in a frequency synthesizer. Voltage pulses representing the modulation are applied to the VCO control input, together with the output of the low-pass filter (LPF) following the phase detector, as shown in Figure 5-10. The modulation spectrum must be higher than the cut-off frequency of the filter to prevent the PLL from correcting the output frequency and canceling the modulation. NRZ modulation is not suitable since a string of ones cannot be held at the offset frequency. Manchester or Biphase Mark or Space modulation is appropriate for FSK since its frequency spectrum is well defined and is displaced above one half the bit rate.

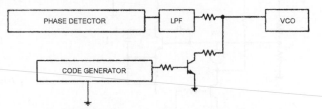

Figure 5-10: FSK With Synthesizer

PSK

As mentioned in Chapter 4, PSK isn't usually used for simple short-range communication systems, particularly because coherent detection is required. Generating the phase-modulated transmitter signal is simple enough and can be done using an IC balanced modulator.

5.3 Amplifiers

As we have seen earlier, the simplest low-power, short-range transmitters may consist of a single oscillator transistor stage. An RF amplifier stage may be added for the following reasons:

- to reduce proximity effects by isolating the oscillator from the antenna

- to make up for losses in a spurious response rejection filter

- to provide for ASK modulation when it is not desirable to modulate the oscillator directly

- to increase power output for greater range

Proper input and output matching is very important for achieving maximum power gain from an amplifier. In order to design the matching circuits, you have to know the transistor input and output impedances. Manufacturer's data sheets give the linear circuit parameters for finding these impedances. However, in many cases, the amplifier stage does not operate in a linear mode so these parameters are not accurate. Trial and error may be used to find good matching, or a nonlinear Spice simulator may be used to determine the matching components. Nonlinear amplification is most efficient but linear amplification is necessary for analog AM systems and digital phase-modulated signals for preventing bandwidth spread due to signal amplitude distortion.

In order to design matching networks, the output impedance of the previous stage, which may be the oscillator, must be known, and also the input impedance of the following stage, which may be a filter or an antenna. Low-power transistor amplifiers typically have low-impedance inputs and high-impedance outputs. RF amplifier output impedance and impedance-matching circuits were discussed in Chapter 3.

5.4 Filtering

In many cases, the maximum power output that can be used in a low-power transmitter is determined by the spurious radiation and not by the maximum power specification. For example, the European specification EN300 220 allows 10 milliwatts output power but only 250 nanowatt harmonics, a difference of 46 dB. The harmonics radiated directly from an oscillator will rarely be less than 20 dB so an output filter, before the antenna, must be used in order to achieve the full allowed power.

An effective filter for a transmitter is a SAW-coupled filter. It has a wide enough pass-band and very steep sides. However, to be effective, the oscillator must be shielded to prevent its radiation from bypassing the filter. The filter has an insertion loss of several dB. It has relatively high input and output impedances and can be matched to a low impedance, usually 50 ohms, by the circuit shown in Figure 5-11. Values for the matching circuit components are given by the device manufacturer.

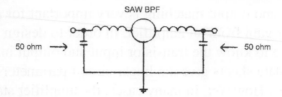

Figure 5-11: SAW Bandpass Filter Matching

5.5 Antenna

We reviewed the theory of various antennas used in short-range wireless operations in Chapter 3. Here we discuss practical implications of their use.

Probably the most popular antenna for small transmitters used in security and medical systems is the loop antenna, discussed in Chapter 3. It's compact, cheap and relatively nondirectional. A typical matching circuit for a loop antenna is shown in Figure 5-12. Resonant capacitance for a given size loop is calculated in the Loop Antenna worksheet. C1, C2 and C3 contribute to this capacitance. Some trial and error is needed to determine their values. C1 should be around 1 or 2 pF at UHF frequencies if the output transistor is the oscillator, to reduce detuning. Making C3 variable allows adjustment of the transmitter to its nominal frequency.

The loop is sensitive to hand effect and, when it is printed on the circuit board, installing the product with the board parallel to and within a few millimeters of a metal surface can both pull the frequency (even if a buffer stage is used) and greatly reduce radiated power. A more effective loop antenna can be made from a rigid wire loop supported above the circuit board.

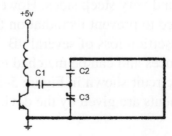

Figure 5-12: Loop Antenna Matching Circuit

Helical and patch antennas are described in Chapter 3. The helical antenna has higher efficiency than the loop and doesn't take up printed circuit board space, but it's also affected more by the hand effect and therefore may not be suitable for handheld devices. The patch antenna is practical for the 900-MHz ISM band and higher, although variations such as the quarter-wave patch, which has one edge grounded, and the trapezoid patch, resonate with smaller dimensions and may even be considered for 433 MHz.

Other antenna forms are variations on monopole and dipole antennas and involve printing the antenna element or elements in irregular shapes around the circuit board. Inductive matching components are needed to resonate the antenna. Antenna performance is affected by proximity of circuit printed conductors and components, including batteries.

While most of the reduced-size antennas are very inefficient, that fact alone doesn't affect the range of the product, since the specifications limit radiated power, not generated power. Thus, the transmitter can be designed to have excess power to make up for the loss of power in the antenna. However, antenna directivity in three dimensions is a limiting factor for products whose position in respect to the receiver cannot be controlled. Another problem is spurious radiation, which is more difficult to control when an inefficient antenna is used.

Detailed descriptions of compact antennas for low-power transmitters are given in application notes from RF Monolithics and from Micrel, referenced in the bibliography.

5.6 Summary

In this chapter we described the basic transmitter building blocks—the oscillator, modulator, RF amplifier and antenna. Size, portability, power consumption and cost are often the determining factors when deciding between different alternatives. Just as in other electronic devices, increasing the scale and functionality of integrated circuits, as well as advances in hybrid technologies to put RF circuits on the same chip with logic elements means that higher performance is possible while staying within size and cost restraints. We'll examine modular short-range transmitter implementations in Chapter 8.

Receivers

It is the receiver that ultimately determines the performance of the wireless link. Given a particular transmitter power, which is limited by the regulations, the range of the link will depend on the sensitivity of the receiver, which is not legally constrained. Of course, not all applications require optimization of range, as some are meant to operate at short distances of several meters or even centimeters. In these cases simplicity, size, and cost are the primary considerations. By far most short-range receiver designs today use the superheterodyne topology, which was invented in the 1920s. However, anyone who starts designing a short-range wireless system should be aware of the other possibilities so that an optimum choice can be made for a particular application.

6.1 Tuned Radio Frequency (TRF)

The tuned radio frequency receiver is the simplest type conceptually. A diagram of it is shown in Figure 6-1. The antenna is followed by a band-pass filter or turning circuit for the input RF frequency. The signal is amplified by a high-gain RF amplifier and then detected, typically by a diode detector. The digital baseband signal is reconstructed by a comparator circuit.

The gain of the RF amplifier is restricted because very high gain in a UHF amplifier would be subject to positive feedback and oscillation due to the relatively low impedance of parasitic capacitances at RF frequencies. An advantage of this type of receiver is that it doesn't have a local oscillator or any other radiating source so it causes no interference and doesn't require FCC or European-type approval.

A clever variation of the TRF receiver is the ASH receiver developed by RFM. A basic block diagram is shown in Figure 6-2.

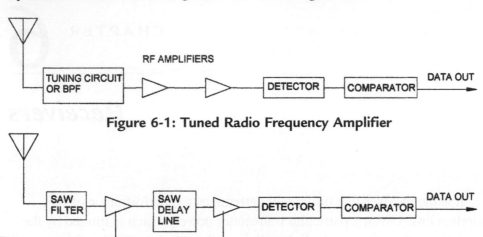

Figure 6-1: Tuned Radio Frequency Amplifier

Figure 6-2: ASH Receiver

In the ASH (Amplifier-Sequenced Hybrid) receiver, two separate RF gain stages are used, each of them switched on at a different time. The gain of each stage is restricted to prevent positive feedback as mentioned above, but since the incoming signal goes through both stages, the total effective gain is the sum of the gains (in decibels) of each stage. While one stage is on the other is off, and the signal that was amplified in the first stage has to be retained in a delay or temporary storage element until the first stage is turned off and the second stage is turned on. The delay element is a SAW filter with a delay of approximately .5 μs.

The selectivity of the ASH receiver is determined principally by the SAW bandpass filter that precedes the first RF amplifier and to a lesser degree by the SAW delay line.

The TRF receiver is a good choice for wireless communication using ASK at distances of several meters, which is adequate for a wireless computer mouse or other very short-range control device. Longer ranges are achieved by the ASH design, whose sensitivity is as good as the best superregenerative receivers and approaches that of superhet receivers using SAW-controlled local oscillators. The TRF and ASH receivers have low current consumption, on the order of 3 to 5 mA, and are used in battery operated transceivers where the average supply current can be reduced even more using a very low-current sleep mode and periodic wake up to check if a signal is being received.

6.2 Superregenerative Receiver

For many years the most widely used receiver type for garage door openers and security systems was the superregenerative receiver. It has relatively high sensitivity, is very inexpensive and has a minimum number of components. Figure 6-3 is a schematic diagram of such a receiver.

The high sensitivity of the superregenerative receiver is obtained by creating a negative resistance to cancel the losses in the input tuned circuit and thereby increasing its Q factor. This is done by introducing positive feedback around the input stage and bringing it to the verge of oscillation. Then the oscillation is quenched, and the cycle of oscillation buildup and quenching starts again. The process of introducing positive feedback and then quenching it is controlled by periodically altering the bias voltage on the transistor input stage, either by using a separate low-frequency oscillator or by dynamic effects within the input stage itself (self-quenched).

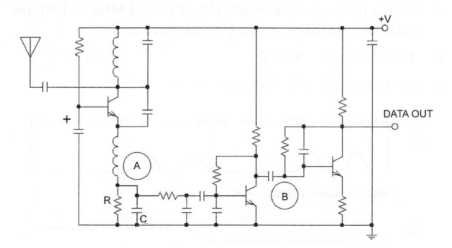

Figure 6-3: Superregenerative Receiver

Figure 6-4 shows the principle of operation of the superregenerative receiver. The two curves refer to points A and B on the schematic in Figure 6-3. Like any oscillator when it is turned on, random noise starts the build-up of oscillations when there is no signal, shown as "0" in the diagram. After the oscillations have been established, the circuit reaches cutoff bias and the oscillations stop, until the conditions for positive feed-back are reinstated. This build-up and cut-off cycling continues at a rate of between 100 and 500 kHz. When a signal is present, the build-up starts a little bit early, as shown as "1" in the diagram. In the self-quenched circuit, the area under each pulse is the same when a signal is present and when it is not, but the earlier starting of the oscillations raises the pulse rate. Averaging the oscillation pulse train by passing it through an integra-tor or low-pass filter results in a higher DC level when a "1" is received, because there is more area under the envelope per unit time than when no signal is present. This signal is amplified by a high-gain baseband ampli-fier and the output is the transmitted data.

In spite of its simplicity and good sensitivity, the superregenerative receiver has been largely replaced because of the availability of inexpen-sive superheterodyne receiver chips. Its disadvantages are

- it re-radiates broad-band noise centered on its nominal receiving frequency;

- it has a relatively broad bandwidth of several MHz on UHF fre-quencies, and thus is sensitive to interference;

- dynamic range is limited;

- it is usable only with ASK modulation.

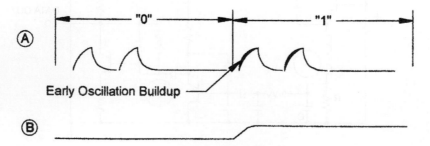

Figure 6-4: Superregenerative Receiver Operation

6.3 Superheterodyne Receiver

The superheterodyne receiver is the most common configuration for radio communication. Its basic principle of operation is the transfer of all received channels to an intermediate frequency band where the weak input signal is amplified before being applied to a detector. The high performance of the receiver is due to the fact that amplification and filtering of input signals is done at one or more frequencies that do not change with the input tuning of the receiver and at the lower intermediate frequency, greater amplification can be used without causing instability.

Figure 6-5 shows the basic construction of a superheterodyne receiver. The antenna is followed by a band-pass filter that passes all signals within the tuning range of the receiver. The mixer multiplies the RF signals by a tunable signal from a local oscillator and outputs the sum and the difference of f_{rf} and f_{osc}. The IF amplifier amplifies and filters either the difference or the sum signal and rejects the other. In VHF and higher frequency receivers it is the lower, difference frequency signal that is retained almost universally and this is the case we consider here. The result is an intermediate frequency signal that has all of the characteristics of the input RF signal except for the shift in frequency. This signal can now be demodulated in the detector. Any type of modulation may be used. A low-pass filter that gives additional noise reduction follows the detector. In a digital data receiver a signal-conditioning stage converts the baseband receive signal to binary levels for digital signals. The signal conditioner in an analog receiver is often a signal expander for audio analog signals that were compressed in the transmitter to improve their dynamic range.

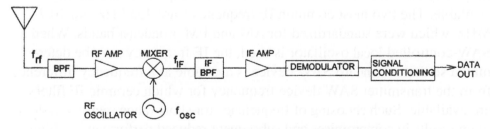

Figure 6-5: Superheterodyne Receiver

The diagram shows only the basics of the superheterodyne principle, but variations are possible for improved performance. The IF bandpass filter (BPF) should be as narrow as possible in order to reduce the noise without affecting the bandwidth required by the modulation components of the signal. The lower the IF frequency, the narrower the bandpass filter can be. However, using a low IF frequency means that the oscillator frequency must be close to the received RF frequency. Since the IF frequency is the absolute value of the difference between the received frequency and the oscillator frequency, two received frequencies can give the same IF frequency—one at $f_{osc} + f_{IF}$ and the other at $f_{osc} - f_{IF}$. It is the function of the input BPF to reject the undesired frequency, which is called the image frequency. When a low IF is used in order to obtain a narrow passband, the input BPF may not reject the image frequency, thereby increasing the possibility of interference. Even if there is no interfering signal on the image frequency, the noise at this frequency will get through and reduce the signal-to-noise ratio. To reduce input noise at the image frequency, including the circuit noise of the RF amplifier, an image frequency bandpass filter should be included between the RF amplifier and the mixer (not shown in the diagram).

In order to reduce the response to image frequencies and have a low IF for effective filtering, UHF superheterodyne receivers often employ dual or triple conversion. A dual conversion superhet receiver has a first mixer and an IF high enough to reject the image frequency using a simple bandpass filter, followed by another mixer and IF at a low frequency for effective filtering.

IF frequency must be chosen according to the image rejection and filtering considerations discussed above, but the final choice usually is a frequency for which standard bandpass filter components are readily available. The two most common IF frequencies are 455 kHz and 10.7 MHz, which were standardized for AM and FM broadcast bands. When a SAW-controlled local oscillator is used, the IF frequency may be determined so that a standard SAW device is available at a frequency difference from the transmitter SAW device frequency for which ceramic IF filters are available. Such choosing of frequencies for short-range security systems often results in compromises and subsequent reduced performance—high image frequency response and insufficient IF filtering.

6.4 Direct Conversion Receiver

The direct conversion receiver is similar to the superhet in that a local oscillator and mixer are used, but in this case the IF frequency is zero. The image frequency, a potential problem in the superhet, coincides with the desired signal, so it is no issue in this topography. Very high-gain baseband amplification is used, and a baseband low-pass filter achieves high sensitivity with high noise and adjacent channel interference rejection.

On the negative side, the local oscillator is at the same frequency as the received signal, so there is a potential for self-interference, and interference with close-by receivers tuned to the same frequency. Design and layout are very important to limit radiation from the local oscillator and prevent leakage back through the mixer and RF amplifier to the antenna. Also, because of the very narrow bandwidth, the crystal-controlled local oscillator frequency must be accurate and stable.

A block diagram of an FSK direct conversion data receiver is shown in Figure 6-6. The output of an RF amplifier is applied to two mixers. A local oscillator output at the same frequency as the RF signal is applied directly to one mixer. The other mixer receives the local oscillator signal after being shifted in phase by 90 degrees. The outputs of the mixers are each passed through low-pass filters and limiters and then applied to a phase detector for demodulation. The frequency shift keyed signals appear with opposed relative phase at the phase detector, giving a binary mark or

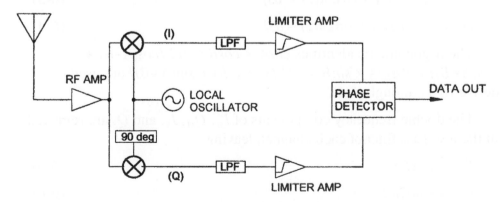

Figure 6-6: Direct Conversion Receiver

space output according to whether the input signal is higher or lower than the local oscillator frequency. This can be seen as follows:

Let input MARK and SPACE signals be

$$S_M = cos2\pi(f+d)t$$

$$S_S = cos2\pi(f-d)t$$

where f is the nominal receiver frequency and d is the frequency deviation of the FSK signal.

The quadrature oscillator signals to the mixers are

$$LO_I = cos2\pi f$$

$$LO_Q = sin2\pi f$$

The mixer outputs when a MARK is sent are:

$$I_M = S_M \times LO_I$$

$$= cos2\pi(f+d)t \times cos2\pi f$$

$$= \frac{1}{2} \{cos2\pi d + cos2\pi(2f + d)\} \tag{6-1}$$

$$Q_M = S_M \times LO_Q$$

$$= cos2\pi(f+d)t \times sin2\pi f$$

$$= \frac{1}{2} \{-sin2\pi d + sin2\pi(2f + d)\} \tag{6-2}$$

Similarly when a SPACE is sent

$$I_S = \frac{1}{2} \{cos2\pi d + cos2\pi(2f + d)\} \tag{6-3}$$

$$Q_S = \frac{1}{2} \{sin2\pi d + sin2\pi(2f - d)\} \tag{6-4}$$

The trigonometric identities $cosA \times cosB = 1/2 \{cos(A-B) + cos(A+B)\}$ and $sinA \times sinB = 1/2 \{sin(A-B) + sin(A+B)\}$ are used to derive (6-1) through (6-4).

The double frequency components of I_M, Q_M, I_S, and Q_S are removed in the low-pass filter of each channel, leaving

$$I'_M = cos2\pi d \tag{6-5}$$

$$Q'_M = -sin2\pi d \tag{6-6}$$

$$I'_S = cos2\pi d \tag{6-7}$$

$$Q'_S = sin2\pi d \qquad (6\text{-}8)$$

where the multiplying constants, 1/2, have been left out.

The limiter amplifiers square up the filter outputs so that they can be applied to a digital phase detector, which outputs MARK or SPACE according to the phase difference in each pair of inphase (I) and quadrature (Q) signals.

Although in the above explanation, the local oscillator frequency is set exactly to the nominal transmitter frequency, small differences are tolerated as long as they are less than the modulation deviation minus the data bit rate.

6.5 Digital Receivers

Digital receivers are more often called "software-defined radios" (SDR). Digital signal processing (DSP) components are extensively used in them and many performance characteristics are determined by software. The basic construction is superheterodyne, but the mixing and IF filtering is done by analog-to-digital converters and digital filters. Although not presently used in short-range products of the types we're focusing on here, they can be expected to be applied to them as costs come down and demands for uncompromising wireless performance and compliance with multiple standards in a single radio force a break from the traditional topologies.

Figure 6-7 shows a diagram of a digital receiver and transmitter. As in a conventional superhet, the signals from the antenna are amplified by a

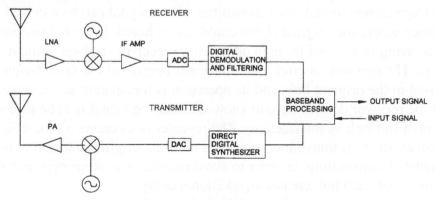

Figure 6-7: Software Radio

low-noise amplifier, then down-converted by mixer and local oscillator to an intermediate frequency. An analog-to-digital converter (ADC) replaces the second mixer of a double superhet, and digital signal processing software performs the IF filtering and demodulation.

The digital transmitter uses a direct digital synthesizer (DDS) to generate the modulated transmitter frequency. This device outputs digital words which represent the waveform of the signal to be transmitted. Phase, frequency and amplitude variations of the carrier as functions of the baseband data are implemented by the software, and many modulation formats can be used as required with no change in hardware. The digital output of the DDS is converted to an analog signal by a digital-to-analog converter (DAC), up-converted by local oscillator and mixer (remember, the mixer also outputs the sum of the input and local oscillator frequencies) and amplified, then coupled to the antenna.

By performing modulation and demodulation in software, and filtering too, great flexibility is achieved, as well as high performance. Elimination of many hardware components will eventually reduce size and cost, while giving high communication efficiency in bandwidth utilization and error correction.

6.6 Repeaters

While range and communication link reliability are limited in unlicensed devices by the low power allowed by the telecommunication authority standards, the use of repeaters can overcome these limitations. In the repeater, a weak signal is received and demodulated. The re-created baseband signal then modulates a transmitter whose signal can be received at a distance where the original signal could not be heard. In digital systems, the relaying of a signal through one or more receivers is done without errors. The repeater receiver and transmitter operate at the same frequency as used in the original link, and its operation is transparent in the sense that the receiver doesn't have to know whether the signal is to be received direct or through an intermediary. The repeater does create a time delay, however, since its transmitter must wait until the original transmitter has completed transmitting, in order to avoid interference. Repeaters may be chained, but each link creates an additional delay.

A potential problem when more than one repeater is deployed is that a repeater closer to the transmitter may repeat the transmission from a repeater further along the link, thereby causing a ping-pong effect. One way to avoid this is to include in the message protocol identification of the repeaters so that a repeater will ignore messages received from a device further down the link. Another way is to force a time delay after transmission of a frame during which an identical frame (received from another repeater) will not be retransmitted.

6.7 Summary

Most often, short-range radio link performance is determined primarily by the receiver. This chapter has reviewed various topologies that have been developed for different performance levels and applications. While the superheterodyne receiver is dominant, the simplicity and low cost, as well as low power consumption, of tuned radio frequency and superregenerative receivers can be taken advantage of for very short-range applications—up to several meters. Digital, or software, radios were introduced here in anticipation of their adoption for short-range applications in the future. Finally, we described how repeaters can be used to extend the range of low-power license-free communication links.

7

Radio System Design

Short-wave radio systems have suffered a bad reputation because of inferior performance and low reliability as a result of poor design. The fact that some of the applications seem deceptively simple, along with the pressures for very low cost, are partially responsible for this. Even in applications where wireless seems obviously more appropriate than wired connections, the market potential has not been realized. While the availability of improved components, such as SAW resonators and advanced integrated RF circuits, has raised the quality and reliability of short-range systems, they do not obviate the need for good design in order to achieve the best performance.

The key to proper design of a short-range wireless system is to match the design to the application. SAW oscillators improved reliability of wireless keys and remote control devices, but in many security systems their frequency accuracy is not sufficient to attain the narrow bandwidth needed for maximum range, given the low power allowed by unlicensed regulations. Current consumption is also an important consideration for many portable battery-operated applications, and compromises in range and spurious response may be necessary to achieve maximum life from small batteries.

The aim of this chapter is to describe the main parameters used to define the performance of wireless receivers and transmitters. We give some examples of needed calculations and discuss tradeoffs that may be considered in certain applications. Most of the discussion is centered on digital systems but analog applications are also referred to.

7.1 Range

Communication range is probably the most obvious characteristic of a wireless link, and of most interest to the system user. However, it is also the most difficult parameter to specify. We learned in Chapter 2 about the vagaries of radio propagation, about multipath and intersymbol interference. There is also the effect of interfering stations and man-made noise, which we can't predict in advance, particularly on the unlicensed frequencies. The communications path between wireless terminals designed for indoor use can be affected by an infinite variety of obstructions and building materials, radiating devices and mounting heights, and antenna orientations. So to specify a minimum or "typical" range on a data sheet is close to meaningless.

One common way to try to get around the dilemma of specifying the operating range of a wireless system is to specify the "open field" range. However, for this specification to have meaning, say for comparing the capabilities of different products for a similar application, the open field range specification must be accompanied by at least the following testing conditions:

- Height above ground of transmitter and receiver

- Orientation of transmitter and receiver towards each other

- Type of antenna used, if there is a choice

- Whether one or both devices are handheld

- Whether the definition is "best case" or "worse case"—that is, whether nulls of no reception up to the stated range are allowed, or whether the range means reliable reception at *any* distance up to the stated range (see Figure 2-2 in Chapter 2)

- Criteria for successful communication—for example, number of correct messages received per number of messages sent.

Rarely are these minimum conditions given when range is stated on a data sheet, so often an inferior device has a superior data sheet range, even "open field," than better equipment that was tested under more tightly controlled and repeatable conditions.

Even though adherence to the conditions described above in specifying range can give some basis for comparison, there are other factors which affect radio range that are almost impossible to control outside of an electromagnetic radiation testing laboratory: ground conductivity, reflecting objects in near vicinity, incidental transmissions, and noise on the communication channel.

So, while the desired criterion for our wireless equipment is very elusive indeed, we should make use of well-defined and measurable indications of performance when designing the system in order to achieve optimum results, given known and thought-out compromises.

7.2 Sensitivity

Sensitivity is the signal power at the input of a receiver that results in a stated signal-to-noise ratio or bit error rate at the output. It depends on the thermal noise power from the input resistance to the receiver, the internal noise generated in the receiver expressed as noise figure, the noise bandwidth of the receiver, and the required output signal-to-noise ratio. We express this sensitivity as follows:

$$(P_{min})_{dBm} = (N_{in})_{dBm} + (NF)_{dB} + (10 \log B)_{dB} + (\text{Predetection S/N})_{dB}$$

$$(7\text{-}1)$$

Let's look at this expression, term by term.

Sensitivity P_{min}

This is the minimum signal power applied to the receiver input terminals that gives the required output signal-to-noise ratio.

Input noise power density N_{in}

The input noise power is the noise originating on the source resistance feeding the receiver, which can be considered as the equivalent resistance of the antenna, or the input resistance of a signal generator used for laboratory testing. The *available* input noise power density, N_{in}, is the maximum power in a bandwidth of one Hertz that this noise source can transfer to a load, achieved when the load resistance equals the source resistance. (When the source has a complex internal impedance, maximum power is delivered when the load is the complex conjugate of the source impedance.)

$$N_{in} = kT \text{ watts per Hertz}$$

where k is Boltzman's constant = 1.38×10^{-23} joules/deg K, and $T = 290$K, the temperature in degrees Kelvin.

Note that the available noise power in a resistor depends only on temperature and bandwidth, and not on the value of the resistance.

N_{in} is in watts per Hertz, and expressing it in dBm for (7-1)

$$(N_{in})_{dBm} = -174 \text{ dBm}$$

This is the ultimate sensitivity of a perfect receiver at room temperature having a bandwidth of 1 Hz, no internal noise, and ability to operate with a signal having the same power as the input noise.

Noise Figure $(NF)_{dB}$

The noise figure in dB is the difference between the input S/N and output S/N (at room temperature), expressed in dB as

$$(NF)_{dB} = (S/N_{in})_{dB} - (S/N_{out})_{dB}$$

It's a measure of the deterioration of the S/N as a signal passes through a two-port network due to the contribution of internal noise. When expressed as a ratio instead of in dB, the noise figure is called noise factor, F. You can find one from the other from the expression: $(NF)_{dB} = 10 \log F$. Pay particular attention in the following discussion whether reference is to F, noise factor, or NF, noise figure.

The expression $(NF)_{dB}$ of Eq. (7-1) is the overall noise figure of the receiver, calculated from the individual noise figures of function blocks in the receiver amplification chain—amplifiers, filters and attenuators, and mixers. Eq. (7-2) shows this calculation.

When used with an amplifier, the gain and the internal noise power can be seen expressly as

$$F = (N_a + G_a N_i)/G_a N_i$$

where N_a is the excess noise power of the amplifier (internal noise), G_a is the gain of the stage, and N_i is the input noise power.

The noise figure of RF active devices—transistors, mixers, amplifiers—is usually given on the data sheet. Noise figure is dependent on bias conditions and for a bipolar transistor can be roughly approximated by

$$(NF)_{dB} = \log(5 \times I_c \times V_c)$$

where I_c and V_c are the collector current in mA and collector-emitter voltage, respectively.

Noise figure is also dependent on input impedance matching, and lowest noise figure usually does not correspond to the optimum matching required for maximum power gain. Thus, the design process involves tradeoffs between noise performance, gain, stability and requirements for matching other components such as filters. The noise figure of passive two- port devices—filters and attenuators for example—is equal to the insertion loss. Insertion loss of filters is stated on the data sheet. The noise figure of a passive mixer is approximately .5 dB higher than its conversion loss.

To find the noise figure of several cascaded two port devices, we use the following formula:

$$F_{total} = F1 + (F2 - 1)/G1 + (F3 - 1)/(G1 \times G2) \ldots \qquad (7\text{-}2)$$

where $F1$ is the noise factor of the first stage, $F2$ is the noise factor of the second stage, and so on, and $G1$ is the power gain of the first stage, $G2$ is the power gain of the second stage, and so on. Remember that

$$F = 10^{\frac{(NF)dB}{10}}$$

When using the noise figure of a mixer in the expression for overall noise figure (Eq. (7-2)), you should note whether the noise figure is specified as single sideband or double sideband. In most VHF and higher frequency communications receivers, the output intermediate frequency is the absolute value of the difference between the input frequency and the local oscillator. Thus, when using a wide-band noise source to measure the noise figure, the mixer will convert the noise source power in the bands both below and above the local oscillator frequency to IF. Most superhet receivers pass the signal through only one of the two frequency bands that the mixer is sensitive to, so it is the single sideband noise figure that we need to use to calculate the cascaded noise figure $(NF)_{dB}$. The mixer single sideband noise figure approaches a limit of 3 dB higher than the double sideband figure, as we see from the following relationship between the single sideband noise factor, F_{SSB}, and the double sideband noise factor, F_{DSB}:

$$F_{SSB} = 2\,F_{DSB} - 1$$

If the receiver front end does not have image suppression we must use the equivalent double sideband noise factor of the mixer and preceeding amplifier, multiplied by two to account for the noise in the image channel. This is the same as doubling the amplifier noise factor by two and using the single sideband noise figure for the mixer. When an image suppressing bandpass filter is not placed at the input to the mixer, but precedes the low noise amplifier (LNA), a composite single sideband noise figure for the mixer and LNA has to be calculated and then used to find the total receiver noise figure. Sometimes more conveniently, the rules which follow can be used to find overall receiver noise figure for the cases where an image rejection filter does not directly preceed the mixer, and where there is no image rejection filter at all, but one or more stages of amplification or attenuation precedes the mixer. An example is given later in this section.

    a. Image rejection filter two or more stages before mixer

- Modify noise factor of stages between filter and mixer, including filter attenuation, by: $F'_i = 2\,F_i - 1$
- Use F_{SSB} for mixer noise factor

    b. No image rejection filter before mixer

- Modify noise factor of stages preceeding mixer by: $F'_i = 2F_i$ (there must be at least one stage)
- Use F_{SSB} for mixer noise factor

Eq. (7-2) shows us that when a preamplifier is used, the noise figure of a chain of devices is influenced predominantly by the first and second stages, since the influence of noise in subsequent stages is reduced by the gains of previous stages. The preamplifier gain should only be enough to reduce the effect of the noise figure of subsequent stages, and no more—excess gain can be detrimental, as we shall see later.

Bandwidth B

Eq. (7-1) shows the importance of narrow bandwidth in attaining high sensitivity. The maximum attainable sensitivity depends on the bandwidth of the modulated signal, since a narrower receiver bandwidth will distort the demodulated signal, causing intersymbol interference and increasing the bit error rate in digital communications and reducing fidelity and intelligibility of analog signals.

In digital systems, the minimum theoretical bandwidth is the Nyquist bandwidth, mentioned earlier, which is one-half of the symbol rate. In a superheterodyne receiver the channel bandwidth is primarily determined by the IF bandpass filter, whose minimum bandwidth in a double sideband receiver is twice the Nyquist bandwidth. Approaching this lower limit of bandwidth is possible only using special high cutoff filters that have a constant delay factor in the passband to prevent distortion. Receivers using these filters must have accurate timing for sampling the data in order to reduce intersymbol interference, which is the blurring of a data symbol due to residual effects from earlier symbols. Such filters are not normally used in simple short-range devices. Instead, discrete component filters make the IF bandwidth much higher than the symbol rate to prevent distortion of the signal.

The signal rates in short-range devices used for control and security may range from several hundred hertz up to around 2 kHz. The bandwidth of an appropriate IF filter could be 10 kHz. However, a great number of these products use SAW resonators for the transmitter and receiver oscillators and standard 10.7 MHz ceramic IF filters with a bandwidth of 300 kHz. Let's assume for the SAW resonators a stability of 20 kHz and an accuracy, after production tuning, of 50 kHz. Then the worst-case frequency divergence of the transmitter and the receiver would be $2 \times (50$ kHz + 20 kHz$) = 140$ kHz. Adding 10 kHz for the modulated signal passband, we see that the receiver IF bandwidth should be at least $(2 \times 140$ kHz$) + 10$ kHz $= 290$ kHz. Thus, the 300 kHz IF filter is justifiable. Referring to Eq. (7-1), the reduction in receiver sensitivity because of the worst-case frequency variation of the SAW resonators is 10 log (290 kHz/ 10 kHz) = 14.6 dB.

The total receiver bandwidth can be less than that determined by the IF bandwidth by using baseband low-pass filtering after the detector, as is done in most receivers. However, baseband filtering is not at all as effective as IF filtering unless synchronous detection is used, or the receiver is a direct conversion type as described in Chapter 6. (In synchronous detection a local oscillator signal having the same frequency and phase-locked to the received signal is multiplied with the received signal recreating, after low-pass filtering, the transmitted baseband data.) The baseband filter will reduce noise, but it will not filter out interfering signals appearing in the IF pass-band, particularly when the desired signal is relatively weak. The nonsynchronous detector is a nonlinear device, and the signal-to-noise ratio at its output is not a linear function of the signal strength. Thus, the band-width of the post-detection filter should not be used for (10 log B) in Eq. (7-1) and its effect should better be expressed in Predetection Signal-to-Noise Ratio, as shown below.

In discussing bandwidth for use in Eq. (7-1), we should define exactly what it means. Commonly, the bandwidth of a bandpass filter is defined as the difference between the upper and lower frequencies where the inser-tion loss increases by 3 dB as compared to the center frequency. The bandwidth of a low-pass filter is the frequency where the response is reduced by 3 dB compared to the DC value. The bandwidth to be used in Eq. (7-2) is the noise equivalent bandwidth (B_N), which is not necessarily the same as the 3-dB bandwidth.

The noise equivalent bandwidth is the bandwidth of an "ideal" perfect rectangular filter that passes the same noise power from a white noise source as the filter whose bandwidth is being determined. White noise has a constant power density over the frequency band of interest. This is exemplified in Figure 7-1, which shows power density spectrums of the two filters. The power spectrum curve of the filter we are interested in is $|H(f)|^2$ and its noise equivalent bandwidth is

$$B_N = \frac{1}{H_0^{\,2}} \cdot \int_0^\infty |H(f)|^2 \, df \qquad (7\text{-}3)$$

where $H_0^{\,2}$ is the maximum power gain of a bandpass filter or DC power gain of a low-pass filter.

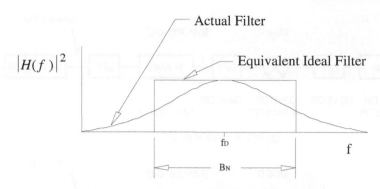

Figure 7-1: Noise Equivalent Bandwidth

For example, the noise equivalent bandwidth of a 3-pole Butterworth type filter is 1.047 times, or around 5% higher, than its 3-dB bandwidth. The noise bandwidth of a single-pole low-pass filter (consisting of a series resistor and a bypass capacitor) is 57% greater than its 3-dB bandwidth. In most cases, except perhaps for the single pole filter, using the known 3-dB bandwidth in Eq. (7-1) gives acceptable accuracy.

Predetection Signal-to-Noise Ratio

The last term in the sensitivity Eq. (7-1) is the signal-to-noise ratio required at the detector input to achieve a defined performance criterion. It may be the signal-to-noise ratio needed for a specified BER, or the specified probability of successfully decoding a message frame. For FM analog detection, the required (signal + noise + distortion)/(noise + distortion) in dB is commonly specified as the performance criterion.

The problematic improvement in sensitivity due to the post-detection filter could be absorbed into the predetection S/N. Thus, effective filtering of the analog detector output before the data slicing or data comparator circuit could reduce the predetection signal-to-noise ratio from, say, 12 dB to 8 dB.

Sensitivity Calculations

Let's now look at an example using the sensitivity equation, Eq. (7-1). Figure 7-2 is a diagram of a superhet receiver, showing the parameters we need to find the sensitivity. Two configurations are shown, differing in the placement of the image rejection bandpass filter.

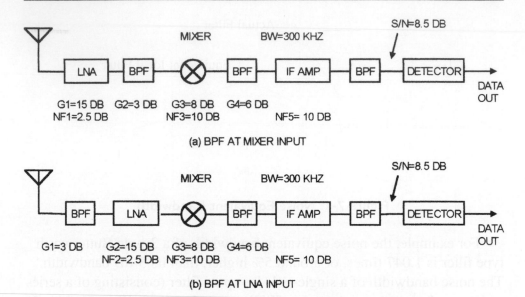

Figure 7-2: Example Receiver for Sensitivity Calculation

$(N_{in})_{dBm}$ is constant and equals –174 dBm/Hz.

The noise figure, calculated up to the IF amplifier, is

$$F = F1 + \frac{F2 - 1}{Ga1} + \frac{F3 - 1}{Ga1 \cdot Ga2} + \frac{F4 - 1}{Ga1 \cdot Ga2 \cdot Ga3} + \frac{F5 - 1}{Ga1 \cdot Ga2 \cdot Ga3 \cdot Ga4}$$

(7-4)

where the Fs are noise factors and the Gas are numerical gains.

The noise figures must be converted to noise factors and gains to ratios, using the relationships:

$$F = 10^{\frac{NF}{10}} \qquad Ga = 10^{\frac{G}{10}}$$

First, we'll find the sensitivity for the configuration shown in Figure 7-2a, where the bandpass (image rejection) filter directly precedes the mixer. The noise factor of the filter has to be modified (see discussion on noise figure above) because its attenuation is effectively after a lossless filter.

Index	Noise Fig. dB (NF)	Noise Factor (F)	Gain dB (G)	Gain Ratio (Ga)
1	2.5	1.78	15	31.62
2	1	1.26 modified>>1.52	–1	.79
3	10	10	8	6.31
4	6	4	–6	.25
5	10	10	—	—

Plugging the *F* and *Ga* columns into Eq. (7-4) we find:

$$F = 2.42 \longrightarrow NF = 10 \log F = 3.8 \text{ dB}$$

The bandwidth term in Eq. (7-1) is

$$10 \log (300,000) = 54.8 \text{ dB Hz}$$

and we will assume that the detector needs an 8.5 dB S/N to output four correct frames out of five, our criterion for minimum sensitivity.

We now find the sensitivity of the receiver to be

$$(P_{min})_{dBm} = -174 + 3.8 + 54.8 + 8.5 = -106.9 \text{ dBm}$$

Now we recompute the sensitivity for Figure 7-2b, where the filter is at the input to the LNA. Here, the noise factors of the LNA and the image rejection filter are modified as explained above.

Index	Noise Fig. dB (NF)	Noise Factor (F)	Gain dB (G)	Gain Ratio (Ga)
1	1	1.26 modified>>1.52	–1	.79
2	2.5	1.78 modified>>2.56	15	31.62
3	10	10	8	6.31
4	6	4	–6	.25
5	10	10	—	—

Plugging the *F* and *Ga* columns into Eq. (7-4) we find:

$$F = 4.1 \longrightarrow NF = 10 \log F = 6.1 \text{ dB}$$

The sensitivity of the receiver for this case is

$$(P_{min})_{dBm} = -174 + 6.1 + 54.8 + 8.5 = -104.6 \text{ dBm}$$

Placing the image rejection bandpass filter at the input to the receiver instead of directly before the mixer causes a degradation of sensitivity of 106.9 − 104.6 = 2.3 dB. However, the configuration of Figure 7.2b is often preferred as it reduces the strength of out-of-band signals that could block the LNA or cause intermodulation distortion.

Calculation of noise figure for cascaded stages is easily carried out using the Mathcad worksheet "Noise Figure." The "Miscellaneous" worksheet has a section for finding sensitivity.

The sensitivity derived from Eq. (7-1) is what would be measured in the laboratory using a signal generator source instead of the antenna. It is useful for comparing different receivers, but it doesn't give the whole picture regarding real life performance.

7.3 Finding Range from Sensitivity

We can tie sensitivity to range with the aid of a plot of path gain vs. distance in an open field. Figure 7-3 is drawn for a frequency of 315 MHz with transmitting and receiving antennas 2 meters above ground, horizontal polarization. The curve shows the ratio, in decibels, of the power at the receiver to the radiated power at the transmitter.

Let's assume that the transmitter is radiating the maximum allowed power. At 315 MHz, a common frequency used in North America, the maximum allowed average field strength per FCC regulations is 6000 μV/m at a distance of 3 meters. The corresponding radiated transmitted power P_t can be found from:

$$P_t = \frac{E^2 \cdot d^2}{30} \tag{7-5}$$

Solving for *E* = 6000 μV/m and *d* = 3 meters, we find P_t = .0108 mW, which is 10 log (.0108) = −19.7 dBm.

We'll assume a receiver antenna gain of –2 dB compared to a dipole, or approximately zero dB isotropic gain. We also assume that the antenna is properly matched to the receiver input impedance.

Now we can find the path gain corresponding to the receiver sensitivity of –106.9 dBm when the receiving antenna has 0 dB gain:

Path gain (dB) = –106.9 dBm – (–19.7 dBm) = – 87.2 dB

From the plot in Figure 7-3 we see that the maximum range of the receiver is 302 m.

This result is greater than what may be expected from the example receiver with the specified field strength, but it doesn't take into account external noise in the frequency band, beyond the thermal noise of the first term on the right of the equal sign in Eq. (7-1). This external noise was discussed in Chapter 2. In order to account for input noise that is different from the room temperature available noise power density, –174 dBm, a couple of modifications are required in Eq. (7-1).

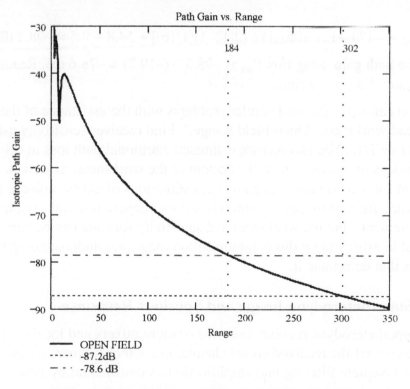

Figure 7-3: Finding Range from Receiver Sensitivity

The input noise, or source noise, can be related to the available noise power by a noise temperature T_s corresponding to noise power density kT_s (k is Boltzman's constant). Then the ratio of the external noise to the room temperature noise = T_s/T_o, where T_o = 290 degrees Kelvin. The revised sensitivity equation to account for external noise different from room temperature resistance noise is

$$(P_{min})_{dBm} = -174 \text{ dBm} + 10\log(T_s/T_o) + 10\log[1+(F-1)(T_o/T_s)] +$$

$$10 \log B + \text{Predetection S/N} \tag{7-1a}$$

The term $10\log(T_s/T_o)$ is the excess external noise in dB, and the third term on the right is a modification of the noise figure (F in the term is the numerical noise figure of the receiver) that is necessary when the input noise temperature is not room temperature.

For an example to show the effect of external noise (man-made noise, for example) on receiver sensitivity, assume the excess noise is 12 dB (ratio of 16)—a fair assumption at 315 MHz. Then, using the same data as before

$$P_{min} = -174 + 12 + 10\log[1 + (2.42-1)(1/16)] + 54.8 + 8.5 = -98.3 \text{ dBm}$$

The path gain using this P_{min} is $-98.3 - (-19.7) = -78.6$ dB. Reading off Figure 7-3 gives a range of 184 m.

You can solve this and similar problems with the assistance of the Mathcad worksheet "Open Field Range." Find receiver sensitivity using Eq. 7-1 or 7-1a. You can include estimated additional path loss in dB as system loss or margin L near the bottom of the worksheet, as well as account for receiver antenna gain. The estimations of excess noise and losses are difficult to make accurately without experience and sample measurements, but the worksheet and sensitivity formula can be very helpful in estimating radio communication range and understanding the factors that determine it.

7.4 Superheterodyne Image and Spurious Response

The superheterodyne receiver uses one or more mixers and local oscillators to convert the received signal channel to another frequency band for more convenient filtering and amplification. A detrimental by-product of this frequency transfer process is the susceptibility of the receiver to unwanted signals on other frequencies. We can see this with the help of Figure 7-4, which shows a dual conversion superheterodyne receiver that has two mixers and two local oscillators.

The desired signal is on 433 MHz which, when combined with the local oscillator at 422.3 MHz in the first mixer, results in a first IF of 433 − 422.3 = 10.7 MHz. After filtering and amplification the signal is applied to a second mixer and a local oscillator at 10.245 MHz, which gives a second IF frequency of 10.7 MHz − 10.245 MHz = 455 kHz. This second IF has the desired narrow passband of ± 5 kHz, which is then amplified and detected.

Why not downconvert directly to 455 kHz and save one local oscillator and mixer? We could do this with a local oscillator of 432.545 MHz or 433.455 MHz, either of which converts 433 MHz to 455 kHz. However, mixing with 432.545 MHz will also convert the undesired frequency of 432.09 MHz to 455 kHz and similarly a 433.455 MHz local oscillator would receive an undesired frequency of 433.455 MHz + 455 kHz = 433.91 MHz. In order to prevent the undesired frequency from reaching the IF filter, it must be filtered out before reaching the mixer. When the undesired frequency is so close to the desired frequency, filtering is difficult if not impossible.

Returning to Figure 7-4, we see that when using a 10.7 IF to precede the 455 kHz IF, the input image frequency is 422.3 − 10.7 = 411.6 MHz, which is practical to separate from the desired 433-MHz signal.

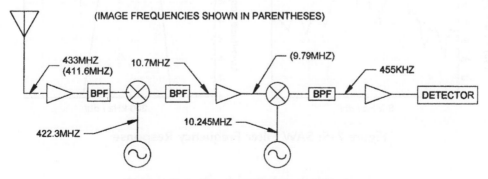

Figure 7-4: Receiver Spurious Response

A second IF could be eliminated by using a 10-kHz bandpass filter at 10.7 MHz. However, such filters are relatively expensive, and a high-gain amplifier at 10.7 MHz with such a narrow bandwidth could easily turn into an oscillator because of stray positive feedback. So, the dual conversion approach is more practical and is widely used.

It is important in designing a superheterodyne receiver to calculate the image responses in order to determine the characteristics of the bandpass filters. In the case of Figure 7-4, a simple LC filter or matching network is not sufficient to eliminate the image frequency, and a helical filter or SAW filter is required at the input. Aside from rejection of a possible interfering signal on the image frequency, this filter reduces unwanted sideband noise before the mixer.

Spurious response can also result from harmonics of strong interfering input signals and of the local oscillator. The following table shows image frequencies and some spurious responses of the receiver of Figure 7-4. Not only must the bandpass filter have high attenuation at the image frequencies and spurious frequencies, but the receiver front-end circuit has to be well laid out and may have to be shielded to prevent very strong interfering signals from getting in directly, not via the antenna, and bypassing the input filter. The curve of a SAW bandpass filter for 433 MHz is shown in Figure 7-5.

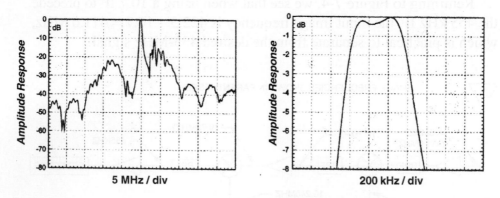

Figure 7-5: SAW Filter Frequency Response

Table 7-1

f_{in} MHz	1st IF MHz	2nd IF kHz	Comments
433	10.7	455	Wanted signal
411.6	10.7	455	1st IF image
432.09	9.79	455	2nd IF image
427.65	10.7	455	Half IF spurious response (2nd Order)
855.3	10.7	455	1st loc. osc. harmonic mixes with input
833.9	10.7	455	1st loc. osc. harmonic image mixes with input
216.5	10.7	455	Sub harmonic of input

The responses shown in Table 7-1 are particular to the mixing process in superheterodyne receivers. Harmonics of the local oscillators and of the input signal are the result of nonlinear effects in the amplifier and mixers, discussed in detail in the next section.

The input RF bandpass filter is very important in reducing spurious responses of superheterodyne receivers. At the least, it must attenuate the image frequency significantly. Other measures must be taken to reduce spurious responses closer to the desired signal frequency.

7.5 Intermodulation Distortion and Dynamic Range

In the previous section we saw that the superheterodyne receiver is susceptible to interference from signals on frequencies other than the one we are trying to receive. In this section we see that new interfering signals can be created in the receiver passband because of the inherent nonlinearity of amplifiers and mixers. Also, strong signals outside the IF band of the receiver can reduce sensitivity to desired signals.

Gain compression

All amplifiers have a limit to the strength of the signal they can amplify with constant gain. As the input signal gets stronger, a point is reached where the output doesn't increase to the same degree as the input; that is, the gain is reduced. A commonly used measure for this property is called the 1 dB compression point. It is shown in Figure 7-6. At the 1 dB point, the output power is 1 dB less, or .8 times, what it would be if the amplifier remained linear. The 1 dB compression point is a measure of the onset of distortion, since an amplified sine wave does not maintain the perfect sine shape of the input. Recalling studies of Fourier series, we know that a distorted sine wave is composed of the fundamental signal plus harmonic signals.

When a strong input signal outside the IF passband reaches the neighborhood of the 1 dB compression point of the input LNA, the gain of the amplifier is reduced along with the amplification of a weak desired signal. This effect reduces sensitivity in the presence of strong interfering signals. Thus, in comparing amplifiers and receivers, a higher 1 dB compression point means a larger capability of receiving weak signals in the presence of strong ones. Note that the listed 1 dB compression point may refer to the input or the output. The output 1 dB point is most commonly given, and it is the input 1 dB point plus the gain in dB.

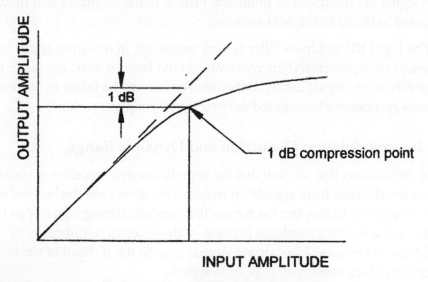

Figure 7-6: 1 dB Compression Point

Third-order intermodulation distortion

A more sensitive indication of signal-handling capability is intermodulation distortion, represented by the third-order intercept. This property is related to the dynamic range of a receiver.

When there are two or more input signals to an amplifier, the existence of harmonics due to distortion creates signals whose frequencies are the sum and difference of the fundamental signals and their harmonics. If there are two input signals, f_1 and f_2, and the distortion creates second harmonics, these are the frequencies of the signals at the output of the amplifier:

$$f_1, f_2, 2f_1, 2f_2, |f_1 - f_2|, 2f_1 - f_2, 2f_2 - f_1$$

If f_1 and f_2 are fairly close to a received frequency f_0 but not in the IF passband, either of the two last terms, $2f_1 - f_2$ and $2f_2 - f_1$, could produce interfering signals closer to f_0. This is demonstrated in Figure 7-7. Both f_1 and f_2 are strong signals outside the IF passband, but they create f_3, which interferes with the desired signal f_0.

The strength of the interfering signal, f_3 in Figure 7-7, increases in proportion to the cube, or third order, of the amplitudes of f_1 and f_2. The intercept point is the point on a plot of amplifier power output vs. input, expressed in decibels, where the power of f_3, the third-order distortion product $2f_1 - f_2$ or $2f_2 - f_1$ at the output of the amplifier, equals the output power that equal amplitude signals f_1 or f_2 would have if the amplification remained linear. Figure 7-8 shows the amplification curves of f_1 or f_2, and f_3, for a 15-dB gain amplifier. The slope of the f_3 curve is three times that of the normal amplification curve. Note that the intercept of the two plots is a graphical point made for convenience. The actual output power in a real amplifier will never reach the intercept point because of the amplitude compression that starts to flatten out the output power at around the 1-dB compression point, which is below the intercept.

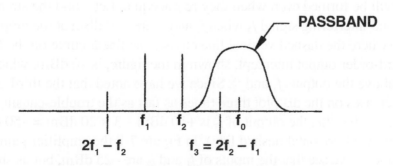

Figure 7-7: Intermodulation Interference

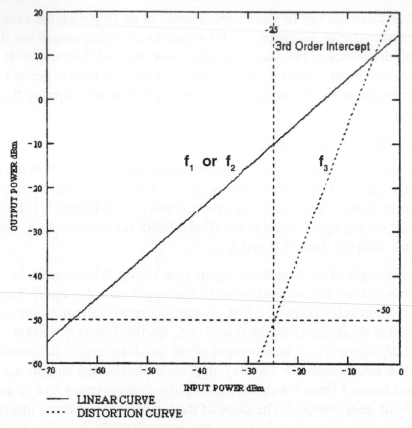

Figure 7-8: 3rd Order Intercept

Although it doesn't actually exist in reality, the intercept point is convenient for finding the spurious signal f_3 for any strength of f_1 and f_2. Again, we take the case where f_1 and f_2 are equal for convenience in measurement and quantizing f_3, although of course the third degree difference signal will be formed even when they're not equal. Let's find the strength of the inband interfering signal f_3 when f_1 and f_2 are –10 dBm at the amplifier output (where the dashed vertical line crosses the linear curve on the figure). The third-order output intercept, shown in the figure, is 10 dBm, which is 20 dB above the outputs f_1 and f_2. Since we have noted that the third-order term increases on the dB plot three times as fast as the trouble-causing signals, we find that the output of f_3 is +10 dBm – 3 × 20 dBm = –50 dBm (marked by a horizontal dashed line). In Figure 7-8 the amplifier gain is 15 dB and so we see that the inputs of f_1 and f_2 are –25 dBm, but as shown in Figure 7-7, outside the receiver IF passband. However, the interference caused by the intermodulation distortion is the same as an equivalent input for f_3, in the receiver passband, of –50 dBm – 15 dBm = –65 dBm. If the data in Figure 7-8 represents the LNA in Figure 7-2, on which we based the sensitivity analysis, and if we are trying to receive a weak signal with this receiver, we can see that the intermodulation distortion in the presence of the interfering signals f_1 and f_2 makes reading the desired signal impossible near, or even tens of dBs above, the calculated sensitivity.

In the example above, we could find the equivalent input power of the "virtual" interfering signal f_3 directly from the distortion curve. Let's now review the steps to analytically find the third-order term, knowing the intercept point and the input power of the distortion-causing signals. Remember that we must know whether the intercept point given for an amplifier represents the input or the output (when given for a complete receiver it must be the input). If the output intercept is given, we can always find the input intercept by subtracting the amplifier gain in dB.

If we don't have the intercept point explicitly from a data sheet, we can use the following rules of thumb to find it and the 1-dB compression point of a bipolar transistor amplifier stage.

Output third-order intercept $OIP3 = 10 \log (V_{ce} \times I_c \times 5)$ dBm

$$(7-6)$$

1 dB Compression $P1_{dB} = OIP3 - 10$ dBm $\qquad (7-7)$

where V_{ce} is the collector to emitter voltage and I_c is the collector current.

We'll demonstrate the process of finding the intermodulation distortion with an example.

Given: Output third-order intercept point $OIP3 = 10$ dBm

Gain $G = 15$ dB

Out-of-band interfering signal input power $P_i = -40$ dBm

Find: Inband equivalent input intermodulation distortion signal power

P_{im}

Step 1: Input intercept point $IIP3 = OIP3 - G = 10 - 15 = -5$ dBm

Step 2: Out-of-band input signal relative to $IIP3 = IIP3 - P_i = -5 - (-40) = 35$ dB down

Step 3: Inband input IM distortion signal

$$P_{im} = IIP3 - (3 \times 35) = -5 - 105 = -110 \text{ dB}$$

Simplifying these steps even more, we find:

$$P_{im} = 3 \times P_i - 2 \times IIP3 = 3 \times (-40) - 2 \times (-5) = -110 \text{ dBm}$$

In the case of the receiver drawn in Figure 7-2, whose sensitivity was found to be -106.3 dBm, the -40 dBm interfering signals will hardly affect weak signal reception.

Second-order intermodulation distortion

In Table 7-1 we saw that our receiver has a spurious response at 427.65 MHz, which is the desired frequency of 433 MHz less one half the first IF—10.7/2. This response is the difference between the second harmonic of the interfering signal—2×427.65—and the second harmonic of the first local oscillator—2×422.3—which results in the first IF of 10.7 MHz. These second harmonics are the result of second-order inter-modulation distortion. This spurious response can be reduced by the image rejection filter between the LNA (Figure 7-4) and the first mixer, which should also reject the harmonics produced in the LNA.

Minimum discernable signal (MDS) and dynamic range

We saw that there is a minimum signal strength that a given receiver can detect, and there is also some upper limit to the strength of signals that the receiver can handle without affecting the sensitivity. The range of the signal-handling capability of the receiver is its dynamic range.

The lowest signal level of interest is not the sensitivity, but rather the noise floor, or minimum discernable signal, MDS, as it is often called. This is the signal power that equals the noise power at the entrance to the demodulator. It can be found through Eq. (7-1), minus the last term, S/N.

The reason for using the MDS and not the sensitivity for defining the lower limit of the dynamic range can be appreciated by realizing that interfering signals smaller than the sensitivity but above the noise floor or MDS, such as those that arise through intermodulation distortion, will prevent the receiver from achieving its best sensitivity.

The upper limit of the dynamic range is commonly taken to be the level of spurious signals that create a third-order interference signal with an equivalent input power equal to the MDS. This is called two-tone dynamic range. It is determined from the input intercept *IIP*3 with the following relation:

$$\text{Dynamic range} = (2/3) \times (IIP3 - MDS) \tag{7-8}$$

While Eq. (7-8) is the definition preferred by high-level technical publications, you will also see articles where dynamic range is used as the difference between the largest wanted signal that can be demodulated correctly and the receiver sensitivity. This definition doesn't account for the effect of spurious responses and is less useful than Eq. (7-8). Another way to define dynamic range is to take the upper limit as the 1-dB compression point and the lower level as the MDS. This dynamic-range definition emphasizes the onset of desensitization.
It may be called single-tone dynamic range.

The two-tone dynamic range of our receiver in Figure 7-2 using an *IIP*3 of –5 dBm and MDS of (–106.3 – 8.5) = –114.8 is, from Eq. (7-8)

$$\text{Dynamic range} = (2/3)(-5 - (-114.8)) = 73.2 \text{ dB}$$

We note here that the intercept point to use in finding dynamic range is not necessarily that of the LNA, since the determining intercept point may be in the following mixer and not in the LNA. The formula for finding the intercept point of three cascaded stages is

$$IIP3 = \cfrac{1}{\cfrac{1}{IIP3_1} + \cfrac{G1}{IIP3_2} + \cfrac{G1 \cdot G2}{IIP3_3}} \tag{7-9}$$

where *G*1 and *G*2 are the numerical gains in of the first two stages and *IIP*3$_1$, *IIP*3$_2$, and *IIP*3$_3$ are the numerical input intercepts expressed in milliwatts of the three stages. Then the total input third-order intercept in dBm will be 10log(*IIP*3).

Note that in calculating IMD of a receiver from its components in the RF chain, you must base the IMD on interfering signals outside of the IF passband and take account of the resulting strength of those interfering signals as they pass through bandpass filters and RF amplifiers.

It should be evident from this description that adding preamplifier gain to improve (reduce) noise figure to increase sensitivity may adversely affect the dynamic range, and using an input attenuator to control intermodulation distortion and compression will raise the noise figure and reduce sensitivity. The design of a receiver front end must take the conflicting consequences of different measures into account in order to arrive at the optimum solution for a particular application.

Automatic gain control

Dynamic range is so important in some short-range radio systems that sometimes special measures are taken to improve it. One of them is the incorporation of automatic gain control, AGC. Figure 7-9 shows its principle of operation.

The incoming signal strength is proportional to the amplitude of the IF signal, which is taken off to a level detector. A low-pass filter averages the level over a time constant considerably larger than the bit time if ASK is used, and longer than short-term radio channel fluctuations if the signal is FSK or PSK. The DC amplifier has a current or voltage output that will control the gain of the LNA and other stages as appropriate to reduce the gain proportionally to the strength of the input signal.

There are some variations to the basic system of Figure 7-9. In one, level detection is taken from the baseband signal after demodulation. For an ASK system, a peak detector with a long time constant compared to the bit time samples the amplitude of demodulated pulses. Also in digital data systems, a keyed AGC implementation is used. Relatively weak signals are unaffected by the AGC, which starts reducing the gain only when the signal strength surpasses a fixed threshold point. Instead of regulating the gain continuously, the gain may be switched to a lower value after the threshold is reached, instead of being adjusted continuously.

While AGC is effective in increasing the single-tone dynamic range, it will not improve two-tone dynamic range when the intermodulation distortion is mainly contributed by the LNA.

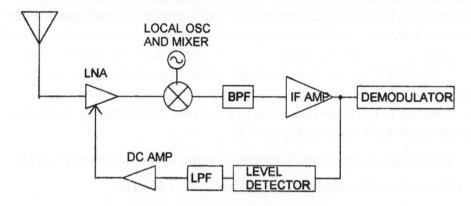

Figure 7-9: Automatic Gain Control

7.6 Demodulation

While the RF and IF sections of the receiver are common to a wide range of system types, the demodulator is specific to the type of modulation used. We will consider the three basic digital data modulation methods—ASK, FSK, and PSK.

ASK

ASK can be demodulated very simply using a diode detector at the output of the IF amplifier. Another way is to use the received signal strength indicator (RSSI) provided in many integrated circuit receiver ICs. This output is a DC current or voltage that is a logarithmic function of the signal strength. The RSSI output facilitates adding AGC, and when used in cellular radios can be part of the transmitter power control function. An RSSI current output must be shunted by a resistor and capacitor for current-to-voltage conversion and filtering before going to the data comparator. The time constant of the resistor-capacitor combination should be less than 10 percent of the width of the narrowest data pulse to avoid unreasonable distortion.

FSK

On receiver and IF circuit ICs, the most common FM or FSK demodulator is a quadrature detector. This is essentially a multiplier that acts as a phase comparator. The amplitude limited IF signal is applied to both inputs of the detector, but a phase difference dependent on frequency is created on one of its inputs by a capacitance in series with a tuned circuit, as shown in Figure 7-10. Figure 7-11 shows how the shape of the output curve depends on the quality factor Q of the tuned circuit. The demodulated output voltage is proportional to the phase difference of the inputs to the multiplier, which is the ordinate of the curve in Figure 7-11.

Determine the required Q as demonstrated in the following example:

Given: Maximum frequency deviation = 40 kHz

Worst-case frequency drift = ±15 kHz

IF = 10.7 MHz

(a) Set D_f to the sum of the maximum frequency deviation and worst-case drift:

$\Delta f = 40 + 15 = 55$ kH

(b) Get maximum normalized IF:

$1 + \text{Max } D_f/\text{IF} = 1 + 55 \text{ kHz}/10.7 \text{ MHz} = 1.005$

(c) Choose relatively straight curve in Figure 11 between normalized IF limits:

$Q = 60$

The resonant circuit components shown in Figure 7-10 can be found using the formulas

$$f_{IF} = \frac{1}{2\pi\sqrt{L \cdot (Cp + Cs)}} \qquad (7\text{-}10)$$

$$Q = \frac{R}{L \cdot 2\pi \cdot f_{IF}} \qquad (7\text{-}11)$$

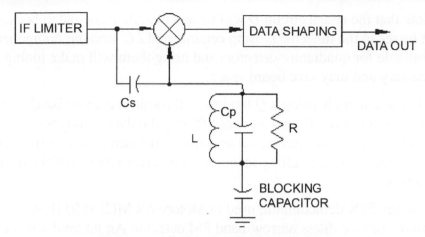

Figure 7-10: FSK Quadrature Detector

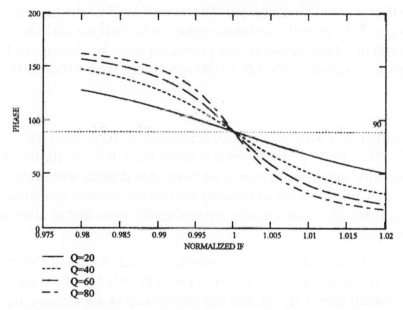

Figure 7-11: FSK Quadrature Detector Frequency Response

Consult the data sheet of the particular IC used to find the recommended *Cs* for the IF used (it may even be included on chip). *R* may also be built in and its value must be used in the expression for *Q* above. For the NE605A IC *R* = 40 kohm and the recommended value of *Cs* is 1 pF. First find *L* using Eq. (7-11) then solve Eq. (7-10) for *Cp* after inserting the calculated *L* and the recommended *Cs* and *R*. Using the above equations for our example, *L* = 9.9 µH and *Cp* = 21.3 pF.

Note that the actual circuit Q will be less than that calculated because of the losses in the inductor L. Both ceramic and LC modular components are available for quadrature detectors and using them will make tuning unnecessary and may save board space.

Also note that choosing a Q too low will result in a lower baseband signal level out of the IC and a lower S/N into the data comparator. Too high a Q can cause unsymmetrical data due to frequency drift across the S curve of Figure 7-11, resulting in an increased error rate out of the data comparator.

Another FSK demodulator, used in Motorola's MC13150 IF subsystem IC, has a coilless narrow-band FM detector. An internal voltage-controlled oscillator connects to one phase comparator input and the limited IF signal to the other input. The demodulator works as a phase lock loop and the demodulated data signal is taken off the oscillator control voltage. Three resistors and a capacitor must be chosen to adjust the loop characteristics. The MC13150 is designed for a 455-kHz IF.

PSK

Phase shift keying is the most efficient modulation form from the point of view of S/N or Eb/N vs. BER, but it is rarely used, if at all, for the short-range security and control systems we have been dealing with. However, it is very often employed for modulating the pseudo-random spreading signal in direct-sequence spread-spectrum applications and in other high data rate systems.

Demodulation of PSK is more complicated than ASK and FSK because it is a coherent process: that is, it needs knowledge of the phase of the transmitted carrier signal, and this information is not directly available. If we input a coherent CW signal representing the RF carrier to one input of a phase comparator and the phase-modulated signal to the other input, the output of the phase comparator will be a baseband signal that reflects the changing phase of the data.

A method for demodulating binary phase shift keying is presented in Figure 7-12. In the squaring loop demodulator shown, the 180-degree phase changes of the modulated carrier are removed when doubling the IF signal input (doubling the frequency doubles the phase, and $2 \times 180 \ \text{deg} = 0 \ \text{deg}$).

This is shown at the top of the figure. A phase-locked loop at the doubled frequency provides a coherent carrier which is divided by two and applied to a phase comparator at the IF. The reconstructed data appears after a low-pass filter.

Another popular BPSK receiving circuit is the Costas demodulator. Its performance is similar to the squaring loop demodulator but frequency doubling is not required. Instead, the input is split and applied to two multiplier elements, whose local oscillator signals are in quadrature. The multiplier outputs, after filtering, are applied to a third multiplier, whose filtered output, free of phase changes, controls the VFO and phase locks its output to the input carrier. Figure 7-13 shows the Costas demodulator.

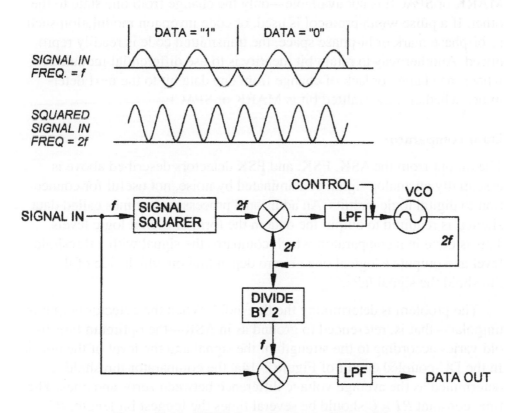

Figure 7-12: BPSK Squaring Loop Demodulator

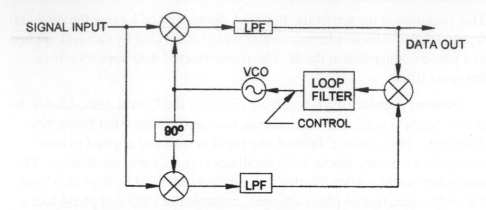

Figure 7-13: BPSK Costas Loop Demodulator

In both types of demodulators, the absolute identity of a data bit as MARK or SPACE is not available—only the change from one state to the other. If a pulse width protocol is used, or code inversion modulation such as bi-phase mark or bi-phase space, the transmitted code is readily reproduced. Another way to retain bit identity is to use differential modulation, where the change or lack of change from one data bit to the next determines whether a transmitted bit is MARK or SPACE.

Data comparator

The output from the ASK, FSK, and PSK detectors described above is essentially an analog signal contaminated by noise, not useful for connection to digital logic circuits. An additional process, sometimes called data slicing, is required to output the data in the form of digital logic levels. This is done in a comparator, which compares the signal with a threshold level and outputs a logical one or zero depending on which side of the threshold the signal falls.

The problem is determining the threshold. When the detector output is unipolar—that is, referenced to ground as in ASK—the optimum threshold varies according to the strength of the signal and the level of the noise. In the DC-coupled circuit of Figure 7-14a, the comparator threshold is determined as the average voltage difference between zeros and ones. The time constant $R1 \times C$ should be several times the longest bit length. $R2$ and $R3$ provide hysteresis, reducing data chatter during bit changes.

Figure 7-14b shows AC coupling via a capacitor. The threshold is derived from a fixed voltage point, in this case a voltage divider. Note that in Figure 7-14(b), as opposed to 7-14(a), the polarity of the output pulses is reversed. AC coupling can be used for baseband data formats that have no DC component, like Manchester coding and bi-phase mark and space.

The capacitors in the data comparators of Figure 7-14 take time to charge up to a new equilibrium after the signal has been absent for some time, or when the receiver is just turned on. Therefore the threshold level isn't optimum during the first few bits received and the error rate may be high. The data protocol should counter this effect by sending a simple preamble of alternating zeros and ones before the actual data, to give time for the threshold to reach its proper value.

a) DC COUPLING

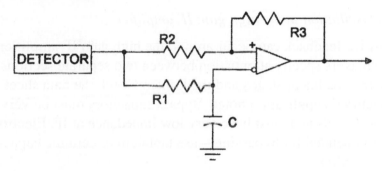

b) AC COUPLING

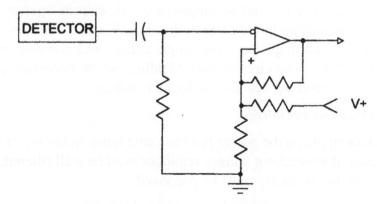

Figure 7-14: Data Comparator

7.7 Internal Receiver Noise

After all measures have been taken to properly design the receiver and prototypes are made and tested, you may discover that the performance falls short of your best intentions. The reason may be interference sources that weren't eliminated completely in the design stage, and trial and error may be necessary to be sure your efforts haven't been in vain.

(a) *Oscillator noise*

A noisy local oscillator can add noise to the mixer output that cannot be eliminated by the IF filter, since most of it occurs in the passband. This noise could be phase noise from the oscillator or spurs in a frequency synthesizer. Changing the oscillator lever to the mixer may help to improve the situation. For a detailed explanation of phase noise see Reference 16.

(b) *Oscillation in the high gain IF amplifier*

Positive feedback can occur around the high gain IF amplifier. Some IF chip makers specify attenuation between two separate sections of this amplifier—the linear stages and the limiter. Check the data sheet and manufacturer's application notes. Bypass capacitors must be very close to the IC terminals and must have very low impedance at IF. Electrolytics are not recommended for bypassing—use tantalum or ceramic capacitors—preferably SMT.

(c) *Radiation from logic circuits*

The microcomputer and accompanying logic circuits operate with switching frequencies in the neighborhood of IF, and harmonics can even reach the received frequency band. Logic circuits and grounds should be isolated from the radio circuits, and shielding may be necessary to prevent reduction of sensitivity because of digital radiation.

(d) *Power supply noise*

Hum or ripple on the power bus can cause noise in the system. Also, the outputs of a switching voltage regulator must be well filtered, and radiation from the supply must be prevented.

7.8 Transmitter Design

There are fewer design topologies in the short-range transmitter than in the receiver, and their influence on overall system performance is less, given a specific allowed power output. However, much more attention must be given in the transmitter to the restrictions placed on unlicensed equipment and failure to recognize and come to grips with potential problems early in the design may result in delays in getting the product to the market. Worse, it may result in a system that doesn't give the performance that even low-power short-range equipment can deliver.

Spurious Radiation

Short-range unlicensed radio regulations, both in North America and in Europe, put more emphasis on reducing radiation outside of the communication channel than on frequency stability and in-channel interference. It's obviously desirable to use the maximum power allowed for best range and link reliability, but this aim is often thwarted because of an excessive level of radiated harmonics whose reduction is accompanied by a reduction of the desired signal power.

It's not the purpose of this section to give detailed instructions on reducing spurious radiation, but rather to call attention to the importance of planning for it early in the design stage. Measures to consider in this regard are

- reduction of printed conductor lengths in the RF stage

- use of small SMT (surface mounted) inductors for low unwanted radiation

- isolating low-frequency circuits and power and battery leads from RF by effective capacitive bypassing and grounding

- providing a high Q output coupling circuit to the antenna

- assuring a good impedance match to the antenna

- choosing and biasing the output transistor for low distortion

- incorporating a bandpass filter between the RF output transistor and the antenna

- shielding the RF stage, or the whole transmitter.

The last two measures usually go together, since they may be the only way to get sufficient isolation between the input of the bandpass filter and the output. Shielding and filters will be necessary to achieve maximum performance for transmitters certified under EN300 220 in Europe, where in most countries 10-mW output power is allowed, but only 250-nW for harmonics, a difference of 46 dB. In the United States and Canada, significantly less power is allowed (except for spread spectrum) under the unlicensed rules, and prudent circuit layout should be adequate to get maximum fundamental radiation while not exceeding spurious radiation limits.

7.9 Bandwidth

As mentioned above, restrictions on bandwidth are not severe for most types of unlicensed operation, a fact that allows the use of convenient and inexpensive SAW-controlled oscillators. However, there are exceptions, and there may be advantages to using narrow bandwidth unlicensed channels where the highest performance and reliability are essential. Examples of narrow bandwidth channels are 40.66 to 40.70 in the U.S. and some places in Europe, 173 to 173.35 MHz in the U.K., and several channels in the European 868-870 MHz band.

The first requirement for meeting narrow bandwidth requirements is a stable frequency source. This can be provided only by a crystal oscillator. Bandwidth occupation depends on the required modulated bandwidth of the transmitted signal plus the stability of the frequency. Narrow-band bandwidth is typically 25 kHz. An FSK signal with data rate of 4 kb/s and deviation of 4.5 kHz will have a minimum bandwidth of approximately 13 kHz. That leaves 12 kHz for accuracy and drift. Thus, the crystal oscillator for 868 MHz must remain within 12 kHz/868 MHz = 14 ppm of the design center frequency.

7.10 Antenna Directivity

Most short-range applications require nondirective antennas. The transmitter or receiver may be portable, and there is usually no way to align transmitting and receiving antennas. Most often the antennas are an integral part of the device and there is no transmission line or connector. During type testing, the maximum radiation is what counts as far as meeting the maximum power requirements, whereas it is the direction of minimum radiation that determines the ultimate performance for random installation orientations. Therefore, an antenna with significant directivity is wasteful of the potential of even the low power allowed for unlicensed devices.

Information on the radiation characteristics of the antennas used on unlicensed devices is given in Chapter 3. Usually there is no defined ground plane on the device, and the antenna is located in close proximity to other components which affect its directivity. Connection wires also have an effect and should be decoupled as well as possible from RF currents. While it is very difficult to predict the radiation pattern of a portable low power device, at least it should be tested and antenna orientation should be varied by trial and error to get a relatively omnidirectional pattern.

7.11 The Power Source

Most short-range devices, and usually transmitters, operate from batteries. The inconvenience of changing batteries is often a deterrent to the use of wireless devices. Security system sensors and portable control devices with very low current consumption rarely use rechargeable batteries. However, wireless telephones and other devices with medium to high current consumption use rechargeable batteries in preference to primary cell batteries, whose periodic replacement would be inconvenient and costly. Security system and other receivers require more energy than intermittently operated transmitters, and they normally have access to a mains power source, but they are almost universally backed up by secondary rechargeable batteries.

The various battery types available certainly affect wireless system design. Power supplies for portable equipment must be designed for specific kinds of batteries. First of all, batteries come in many different voltages. Unlike mains supply, battery terminal voltage and internal resistance change during battery life or discharge time. Requirements for lowest possible current drain also dictate the type of voltage regulation to be used, or often makes the designer do with no regulation at all. Thus, a handheld remote control transmitter may have good range just after the battery is replaced, but performance will often noticeably deteriorate due to reduced transmitter output power as the battery wears out. Many portable wireless devices have low-battery detection systems that indicate either visually or by signaling a control panel that the battery needs replacement. It is important for the wireless system designer to determine the appropriate battery type early in the design stage. We give here a brief description of several battery types used in short-range radio applications. Manufacturer's specifications should be consulted for more details.

Nine-volt alkaline batteries have been very popular for portable control units and security sensors, as well as 12-volt alkaline batteries for wireless keys. Many new devices are designed using lithium batteries. Most of the standard types of active electronic components and integrated circuits operate at 3 volts and below, so the 3-volt lithium cells can be used directly. Incorporation of lithium batteries in portable wireless equipment has extended the replacement period to 5 years and more. Lithium batteries are not homogeneous, and there are several chemical systems in use, each with its own performance and safety characteristics. They're described briefly below [see Reference 35].

Poly (carbon monofluoride) cells have an open-circuit voltage (OCV) of 2.8 volts and moderately high energy density. The cells are available in all standard cylindrical sizes as well as coin types.

Manganese dioxide lithium cells are also available in standard sizes. They have similar characteristics to the poly (carbon monofluoride) type. These cells have lower internal impedance than the other lithium battery types so they are well suited to applications needing relatively high continuous current or pulse requirements. A plot of output voltage vs. discharge time for various loads is given in Figure 7-15 for a Duracell type DL123A size 2/3A cell.

The lithium-iodine technology uses only solid constituents and therefore is very safe. Its separator can heal itself if cracks occur. A drawback is high internal impedance, so lithium-iodine batteries are practical only for very low-current applications. The majority of implanted cardiac pacemakers are powered by lithium-iodine cells. OCV is 2.7 volts.

Thionyl chloride-lithium cells have the highest energy density of all lithium types. They have a service life of 15 to 20 years and are available in all case styles—cylindrical, coin and wafer. The cells are best suited to very low continuous current and moderate pulse demand applications and their long service life and low self-discharge rate make them appropriate in products having limited physical access, such as remote sensing systems. Thionyl chloride-lithium cells have an open circuit voltage of 3.6 volts.

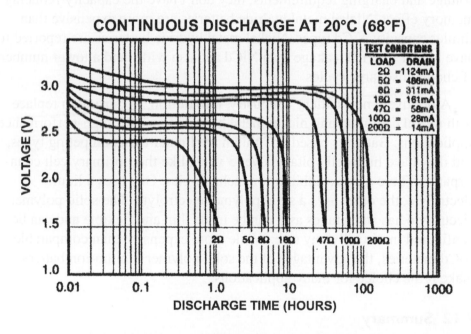

Figure 7-15: Lithium Battery Discharge Curve (Duracell DL 123A)

The staple rechargeable battery type in portable equipment has for many years been nickel cadmium. It has high energy density and tolerance for abuse, as well as being economical. Load voltage of a fully charged NiCd battery is between 1.2 and 1.3 volts per cell. Its terminal voltage is fairly constant until discharge, when it reaches 1 volt per cell. NiCd batteries have very low internal impedance so they can supply high current on demand. A known deficiency of NiCd batteries is the so-called memory effect. Batteries that are recharged often and fully after shallow discharge appear to "remember" their short discharge time and lose capacity. This effect is reversible, and may be corrected by several cycles of complete discharge and recharge.

Nickel metal hydride batteries are taking over many applications of nickel cadmium types. Having similar characteristics, including cell voltage and charging requirements, they don't have the capacity-reducing memory effect. Nickel metal hydride batteries are more expensive than similar capacity NiCd types. Nickel metal hydride batteries are reported to have higher internal leakage than NiCd and can withstand a lower number of charge/discharge cycles.

A relatively new battery chemistry, lithium-ion, is starting to replace both nickel cadmium and nickel metal hydride types for high-performance applications. Batteries based on lithium are lighter than competing types, and they have high cell voltage, over 3 volts, like their primary cell counterparts. There are two lithium-ion technologies. One uses a liquid electrolyte, the other uses a solid polymer electrolyte. The solid polymer electrolyte has advantages as it can be made thin and flexible and can be configured in virtually any size. While more expensive than comparable NiCd batteries, the advantages of the solid polymer lithium-ion batteries make it the choice for many applications.

7.12 Summary

Performance characteristics of a radio communication system are reviewed in some detail in this chapter. We mentioned the conditions that affect communication range, and then dealt with the various factors that determine receiver sensitivity. We saw that preamplification is necessary in a receiver to reduce the effects of internal noise, but only up to a point. Excessive preamplification lowers the threshold of overloading, measured by compression and intermodulation distortion, and reduces the dynamic range. The superheterodyne receiver topology, by far the most common in use, is susceptible to interference by signals on frequencies far removed from the desired frequency to which the receiver is tuned. We saw how to calculate these frequencies and discussed remedies for avoiding them, by filtering, shielding, and circuit layout. Since virtually every modern receiving device has digital logic circuits and microcomputers, the designer must be aware that radiation from those circuits to the receiver front end can reduce sensitivity.

The different modulation methods require different detectors and we discussed the most common ones for short-range communications. The data comparator, or signal slicer, seems to be relatively simple but its design is highly dependent on the baseband modulation format. Faulty design of this component in the receiving link is often responsible for several dBs of lost sensitivity.

Transmitter spurious radiation reduction, as well as attention to the antenna radiation pattern, are keys to obtaining maximum allowed radiated power. Transmitter oscillator stability requirements are closely related to whether narrow-band or broad-band channels are employed.

Finally, we reviewed the characteristics of several portable power sources available, both from one-time-use primary cell batteries and from rechargeable batteries. An important factor in increasing battery life is the reduction of supply voltages for active components. In many applications, a single lithium cell can supply the power requirements of a portable device with no voltage conversion circuits or supply regulation.

8

System Implementation

The designer of a short-range radio system can take any of several approaches to implement their system. This chapter has examples of three levels of component integration for which devices are available from several manufacturers. These levels are:

- *Complete wireless modules.* They comprise a complete system; transmitter, receiver, or transceiver—and are packaged as a small assembled printed circuit board which can be inserted as a plug-in or solder-in component on the user's motherboard, or as a board sub-assembly to be wired to the user's baseband system. The system designer must add the antenna and the baseband application processing. Baseband inputs and outputs are standard digital levels, although some modules accept analog audio signals. There are modules available that have a standard or proprietary data interface, and do the lower protocol processing on board.

- *System on a chip.* These are monolithic or hybrid wireless systems packaged as integrated circuits. They are almost as complete as the modules, but usually need a few more external components, such as SAW or crystal frequency determining oscillator components, bandpass filters, and capacitors and resistors for parameter setting and DC filtering. The receivers amplify and demodulate the desired RF signal frequency and output the data in digital form. Transmitters accept digital or analog baseband inputs and output the RF signal at the required level, although external spurious radiation filtering is usually needed between the RF output and the antenna. Some devices have digital logic blocks or microcontrollers on the chip that perform baseband processing, including digital filtering and receive data clock regeneration.

■ *Large scale wireless integrated circuits.* These include, principally, the IF amplifier chain used in receivers but other blocks such as synthesizers and RF front ends also belong to this category.

There are two other approaches to the design of short-range systems that we do not dwell on here. One is the use of discrete components and small scale ICs. This approach is rarely taken for receivers, except perhaps for lowest-cost, low-performance designs, including those based on superregenerative receivers. Low-cost transmitters for security systems and keyless entry can easily be designed from discrete components and usually are.

At the opposite end of the application spectrum—sophisticated high data rate systems, often using spread-spectrum modulation—some designers will design their own proprietary Application Specific Integrated Circuit (ASIC) based device. This approach is viable for very large production quantities.

Consider the following points before deciding on which approach to take:

■ Overall cost, including development outlays and unit production cost,

■ Time to market,

■ Degree of satisfying particular system requirements with each approach,

■ Availability in-house of RF engineering proficiency and development and production test equipment, and

■ Possible supply problems due to choosing a single source for the main module or IC.

No matter which approach you take, you need basic RF know-how to evaluate and choose the devices to base the design on, develop or adopt appropriate coding protocols, design the antenna, submit the system to testing according to standards and regulations, present the product to the market and back it up for the customers.

Examples of devices from each of the three levels are given below, with comments on their specifications and unique features. An important parameter for choice—price—is not given but a prospective customer should be able to obtain it from the manufacturer or distributor.

Remember that important performance specifications, such as sensitivity and range, should not be taken at face value but should be verified independently before using them to base a decision. While not necessarily the rule, some manufacturers engage in "specmanship," and a claimed high sensitivity may be either obviously impossible or may be stated without giving the error rate or signal-to-noise ratio on which it is based. Range specifications are most often abused, and should never be used for comparison between performance of products of different companies. In considering transmitter power output, remember that the regulations limit radiated power output, and that depends strongly on the antenna design. Declarations regarding strength of harmonics and other spurious radiation from transmitters are generally based on direct measurements at the antenna terminal, whereas their levels can be quite different in radiation measurements on the finished product.

Most of the companies that make modules and system ICs also offer evaluation boards and systems. Trying out a product before committing oneself to it is a good way to reduce the possibility of an expensive mistake.

The inclusion of a product below in no way indicates a recommendation or an endorsement of its characteristics. Many products not listed here may be more suitable for your purposes. All information is based on manufacturers' promotional materials. We tried to be accurate, but specs change, and the manufacturers should be consulted for authoritative information.

8.1 Wireless Modules

Circuit Design, Japan (www.circuitdesign.co.jp)

The company has products for the unlicensed UK and European frequency bands, notably 433 MHz, 458 MHz, and 868 MHz. They are designed to give maximum performance at low data rates and use stable synthesized crystal oscillators and narrow-band receivers. The modules are enclosed in metal cases with pins for soldering to a circuit board.

Example products and features:

CDP-TX-02A Transmitter

- Frequency: 433-MHz band, 32-switch selectable 25 kHz channels

- Modulation: FSK, 2-kHz deviation

- Output power: 10 mW nominal

- Maximum baud rate: 4800 bits/sec

- Frequency control: Crystal-controlled synthesizer

- Stability: 5 ppm

CDP-RX-02A Receiver

- Frequency, stability and demodulation: Compatible with transmitter

- Type: Double conversion superheterodyne

- Sensitivity: –120 dBm for 12 dB SINAD

Radiometrix, UK (www.radiometrix.co.uk)

A range of transmitter, receiver, and transceiver modules, operating on the popular UHF unlicensed bands, is offered by this company. Products differ in case style, performance, and baseband functions. An example is the BiM2-433 transceiver module. It is available for nominal supply voltages of 5 (standard) or 3.3 volts. The module is shielded and can be soldered to or plugged into a motherboard. See Figure 8-1. These are the principle characteristics:

- Nominal frequency 433.92-MHz, SAW controlled

- FSK modulation, maximum data rate 160 Kbps

- Transmitter power output (5 V supply) +10 dBm typical

- Receiver sensitivity –90 dBm at 160 Kbps and BER = 1×10^{-6}

- Carrier detect output

- Logic level and analog receiver outputs

- Built in Tx/Rx antenna switch

- Module dimensions 33mm × 23mm × 4mm

- Tested for conformance with the R&TTE Directive

The BiM2 serves as the RF section of a SpacePort modem that has built-in baseband processing for accepting and outputting data in RS-232 serial format. The device can be configured by the user for a range of host data rates from 600 to 115200 bps and other characteristics including addressing for point-to-multipoint networking and throughput regulation for duty cycle control. The module controller includes transmission framing, error detection, and automatic retransmission of badly received packets. Case dimensions are slightly larger than those of the BiM2-433— 23mm × 39mm × 6.5mm.

Figure 8-1: Radiometrix BiM2 433-MHz Module
(Figure with permission of Radiometrix)

Linx, U.S.A. (www.linxtechnologies.com)

Linx offer several ranges of RF data modules in different cost and performance categories. The company also supplies whip and board mounting helical antennas.

The LC series devices are small, encapsulated modules with SMD packaging designed for automated assembly on product circuit boards. Basic characteristics are:

- ASK modulation

- SAW stabilized transmitters and receivers on standard UHF frequencies

- Superheterodyne receiver

- Data rates under 5000 bps

Also operating on standard UHF frequencies in the 400 MHz range, the RM series are PC board modules with better range than the LC series devices. They also use SAW control but with FSK modulation and data rates up to 10000 bps.

Linx's highest performance devices, the HP series, operate between 902 and 928 MHz. They are PC board modules with synthesized crystal control and eight switchable frequencies. Special features are

- eight switchable frequencies in the 900-MHz band

- Frequency modulation from analogue or digital sources

- Proprietary digital data output circuit in receiver accepts 300 to 50,000 bps with no duty cycle limitations

- Receiver data qualification circuit squelches output when the signal is absent or too weak for accurate demodulation

Adcon Telemetry (Austria) (www.adcon.com)

Adcon's addLINK™ transceiver module for the 868–870 European unlicensed band is distinguished by having a built-in antenna that is part of its packaging, shown in Figure 8-2. The device includes a microcontroller programmed to interface with serial RS-232 data at TTL levels and rates of 1200 to 38400 baud. Other characteristics:

- DC power at 3 to 5 volts

- RF output 0 to 7 dBm

- Center frequency 869.85 MHz

- Sensitivity –101 dBm

- Dimensions 21mm × 38mm × 7mm

The company also offers a line of completely packaged data modems, various transceivers on open PC boards, including spread-spectrum devices for the North American 902–928 MHz band, and custom wireless communication products.

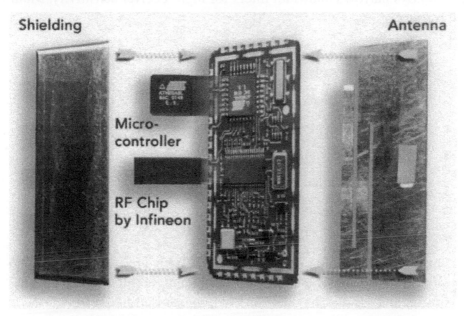

Figure 8-2: Adcon addLINK Transceiver with Built-in Antenna
(Figure with permission of Adcon Telemetry AG)

Honeywell International (www.ssec.honeywell.com)

The company produces "Radio on a Chip" transceivers for the 915 MHz and 2400 MHz ISM bands and modules for 915 MHz. The HRF-ROC09325XMS module comes in a surface-mount package whose size is 23mm square × 4.5mm. Power source voltage may range from 2.8 to 3.3 volts for standby current of less than 5 microamperes. Data and control communication with the device from a host microcontroller is over a three-wire SPI (serial peripheral interface) bus. Some other features are frequency agility, transmitter power up to 6 dBm, and receiver sensitivity better than –70 dBm at 19.2 Kbps.

MK Consultants (UK) (www.mkconsultants.co.uk)

The X2010 transceiver module, pictured in Figure 8-3, can be supplied for use in the 433-MHz, 868-MHz, or 915-MHz bands. It is crystal controlled and employs narrow bandwidth filters for high receiver sensitivity. Main features are:

- Supply voltage, 5 volts
- Data rates up to 10 Kbps
- Receiver sensitivity −112 dBm for 12 dB SINAD on analog output
- Image rejection 50 dB
- Transmitter output power 1 mW
- Frequency modulation
- Digital and analog receiver data or voice outputs
- Carrier detect and RSSI outputs

The company supplies a version, X2011, with a printed loop antenna as part of the module PC board. MK Consultants supplies a range of transmitter and receiver modules on UHF unlicensed frequency bands.

Figure 8-3: MK Consultants X2010 UHF Transceiver
(Figure with permission of MK Consultants)

Bluechip Norway (www.bluechip.no)

Three shielded modules, 25mm square and 3.4mm thick, having identical pinouts, cover the three popular bands of 433 MHz, 868 MHz, and 915 MHz. They are RFB433, RFB868, and RFB915. Packages are surface mountable. Connection to a microcomputer is over five I/O lines, and their frequency synthesizers can be controlled for FHSS. Output power is 10 dBm and receiver sensitivity is −105 dBm on 433 MHz and −104 dBm on the higher bands. Data rate to module is 19.2 Kbps, Manchester encoded, giving a user data rate of 9.6 Kbps.

8.2 Systems on a Chip

Micrel U.S.A. (www.micrel.com)

Micrel offers several variants of an integrated circuit radio receiver which requires no more than an antenna, ceramic resonator, two capacitors and a 5 volt DC source to get on the air. The devices receive ASK and cover the popular 300 to 440, 868, and 900 MHz ranges for unlicensed radio communications. These low-cost units can replace superregenerative receivers while accommodating transmitter frequency drift of ±0.5% but can be configured to give better performance for use with SAW and crystal-controlled transmitters. The receiver is a single-conversion superheterodyne with AGC for good dynamic range. Post-detection data filtering is done on the chip, and filter bandwidth can be selected by the system designer in accordance with the data rate.

Figure 8-4 illustrates the simplicity of the Micrel MICRF001 receiver chip operating at 387 MHz.

Micrel also offers its MICRF500 and MICRF501 transceiver chips for frequency ranges of 700 to 1100 MHz and 300 to 600 MHz, respectively. These devices each have FSK data rates of up to 128 Kbps, and a frequency synthesizer with crystal reference. Antenna switching is external to the chip. Receiver architecture is zero IF type where the channel bandwidth is determined by programmable low-pass filters integrated on the chip.

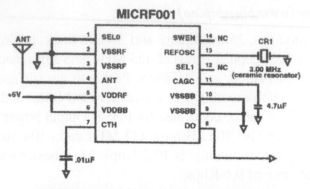

387 MHz, 1200 BAUD OOK RECEIVER

Figure 8-4: Micrel QuikRadio Receiver

RF Monolithics, U.S.A. (www.rfm.com)

In contrast to all the other receivers referred to in this chapter, RFM offers a device which is not a superheterodyne design. The company's patented ASH (amplifier-sequenced hybrid) receiver uses SAW filters to give reasonable selectivity in what is essentially a tuned radio frequency design. More details are in Chapter 6. External connections are shown in Figure 8-5.

Another difference in the RFM ASH receiver product is that the frequency determining elements (the SAW filters—there are no RF oscillators) are included in the SMD package. Since the system designer cannot define the exact frequency by an external resonator (SAW, crystal, or ceramic device in other products), RFM offers a line of products covering virtually all of the commonly used unlicensed UHF frequencies from 303 MHz to 916 MHz. In addition to the frequency, the part number also defines the data rate, up to 115,000 bps.

The ASH design gives the receiver these special characteristics:

- No RF radiation (there are no RF oscillators)

- High interference rejection outside the pass band—no image frequency

- Low operating current—1.1 mA typical from a 3-volt supply

- Few external components

- No tuning

RFM also offers a range of transceivers based on the ASH receiver and a transmitter, both sharing the built-in SAW filter components. The transceiver is contained in a single small SMD package.

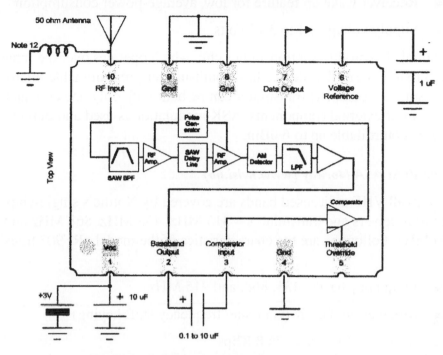

Figure 8-5: RFM ASH Receiver Connections

Atmel Corporation (www.atmel.com)

Atmel's AT86RF211 transceiver chip has high configuration flexibility through external microcomputer control. Its frequency range is 400 to 950 MHz so it is applicable to the most popular unlicensed bands. Main characteristics are:

- Modulation: FSK
- Frequency control: crystal reference synthesizer—supports frequency agility
- Data rate: up to 64 Kbps
- RF power: adjustable up to 14 dBm (on 433 MHz)
- Internal antenna switch

- Receiver architecture: double conversion superheterodyne

- Receiver sensitivity: −105 dBm at 4800 bps, BER = 1%

- Receiver wake-up feature for low, average-power consumption

- Supply voltage 2.4 to 3.75 volts

Atmel also produces a transmitter chip which operates over the range of 264 to 456 MHz. The device has a built-in user programmable micro-processor, so a complete transmitter can be built with only one chip and a minimum of external components. ASK modulation is used and output power is controllable up to 6 dBm.

Nordic VLSI ASA Norway (www.nvlsi.no)

Virtually all VHF unlicensed bands are covered by Nordic's single-chip radio transceivers and transmitters—315 MHz, 433 MHz, 868 MHz and 2400 MHz. Following are the characteristics of the model nRF903 trans-ceiver:

- Frequency bands: 433, 868, and 915 MHz

- Modulation: GFSK (Gaussian frequency shift keying)

- Maximum data rate: 76.8 Kbps

- Oscillator control: Crystal reference frequency synthesizer

- RF power: Maximum 10 dBm

- Antenna switch: Built-in with differential terminals

- Data I/O: Bidirectional data, carrier sense

- Microcomputer control over SPI (Serial peripheral interface)

- Rx architecture: Superheterodyne, image rejection first mixer, second IF 10.7 MHz

- Supply voltage: 2.7 to 3.3 volts

A functional block diagram of the nRF903 is shown in Figure 8-6.

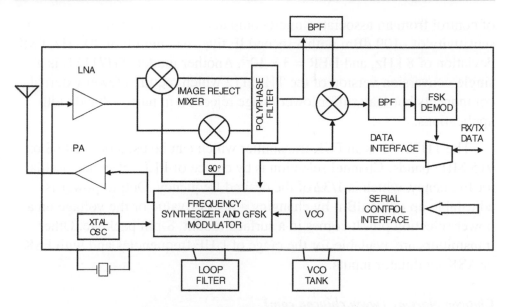

Figure 8-6: Nordic nR903 Transceiver

Nordic's 2.4 GHz offerings include the nRF2401 transceiver and nRF2402 transmitter. They are designed for low-cost, low-current peripheral interconnections and are suitable for FHSS. Data rate is up to 1 Mbps and modulation is GFSK. Reduction of current consumption during transmission is achieved by what the company calls "ShockBurst" technology which clocks in data to be transmitted at a low rate and then sends it in high rate bursts. Another feature is the transceiver's dual receiver topology, which lets it receive simultaneously from two sources on channels separated by 8 MHz, outputting the data on two separate baseband ports.

Melexis Germany (www.melexis.com)

Melexis transmitters, receivers and transceivers operate on the 315, 433, 868 and 915 MHz bands. An example of a Melexis receiver is model TH71112 for 868 or 915 MHz. It uses double-conversion and can demodulate FSK or ASK signals. The local oscillator is based on a PLL with crystal reference and a fixed divider from the internal VFO. Its single operating frequency depends on the frequency of the crystal reference, and, unlike some of the other receivers we have looked at, is independent of control from an associated microcomputer. The manufacturer claims a sensitivity of −109 dBm with a second IF filter bandwidth of 40 kHz, FSK deviation of 8 kHz, and BER = 3×10^{-3}. Another model, TH71111, is a single-conversion version of the TH71112 which requires fewer external components but may suffer lower image rejection in narrow bandwidth applications.

The TH72031 is an FSK transmitter which can be used on the 868 or 915 MHz bands. Channel selection is by choice of PLL reference oscillator frequency, which is 1/32 of the desired frequency. Output power is adjustable up to 8.5 dBm by changing a resistor value or the voltage on a power selection pin. It comes in a surface-mount, 8-pin package. Other transmitters are available for the range of UHF frequencies and with FSK or ASK modulator inputs.

Chipcon Norway (www.chipcon.com)

Chipcon offers a line of transceivers and a transmitter for the unlicensed frequencies between 300 and 1000 MHz. The devices contain a relatively high degree of digital logic for fine frequency control, digital frequency modulation and demodulation, bit synchronization, and transmitter and receiver parameter modification. Figure 8-7 is a block diagram of the model CC1020 transceiver. These are some distinguishing features of selected models:

	CC1000 Transceiver	*CC1010 Transceiver*	*CC1020 Transceiver*	*CC1050 Transmitter*
Distinguishing characteristic	Low-supply voltage and current	Built-in microcontroller core	Narrow-band and high sensitivity	Transmitter only
Supply voltage	2.1 to 3.3V	2.7 to 3.6V	2.3 to 3.6V	2.1 to 3.6V
Supply current	7.4 mA in Rx mode	13.1 mA in Rx mode	17.3 mA in Rx mode	9.1 mA at 0 dBm
Programmable output power	−20 to 10 dBm	−20 to 10 dBm	−20 to 10 dBm	−20 to 12 dBm
Best sensitivity	−110 dBm	−107 dBm	−121 dBm	—

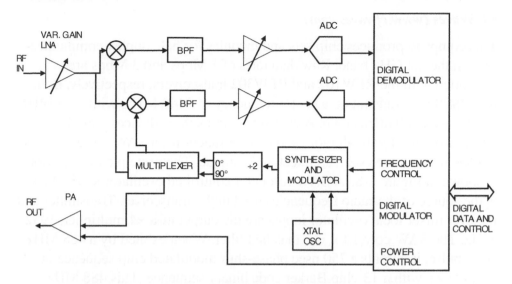

Figure 8-7: Chipcom CC1020 Narrow-Band Transceiver

Xemics Switzerland (www.xemics.com)

Two wireless data communication transceivers, covering the UHF unlicensed bands, are available from Xemics. The XE1201A covers the range of 300 to 500 MHz, for the low UHF North American and European bands. The XE1202 is optimized to operate on the 868 and 915 MHz bands, but can be used on the 433 MHz band as well. Both devices employ a zero IF architecture and use digital FSK modulation and detection. Following are features of the XE1202:

- Sensitivity –116 dBm
- Continuous phase FSK modulation with programmable deviation
- Data rates to 76.8 Kbps
- Allows narrow-band operation for 25 kHz channels
- Frequency synthesizer steps 500 Hz
- Received data pattern recognition
- Received data bit synchronizer
- Programmable output power up to 15 dBm
- Minimum supply voltage 2.4 V

RF Waves (www.rfwaves.com)

This company produces chip sets and modules for half-duplex communication on the 2.4 GHz band. Raw data rates of 1 Mbps and 3 Mbps are supported with the RFW102 and RFW302 transceivers, respectively. Both use DSSS and transmit on a center frequency of 2440 MHz, with a 30-MHz bandwidth at 20 dBm down. A chip set contains three devices—an RF integrated circuit with all radio and control functions, a SAW correlator, and a SAW resonator. System architecture is unique. The principle of operation is shown in Figure 8-8. A SAW resonator controlled oscillator at 488 MHz is the source of the radio frequencies used in the transceiver. Transmitter and receiver mixer local oscillator inputs are the output of a ×4 multiplier, 1952 MHz. The SAW correlator is a matched filter. When excited by a 488 MHz short pulse, it outputs a 750 nsec phase-shift modulated chip sequence in accordance with a 13-chip Barker code binary sequence. This 488 MHz burst is amplified and upconverted in a mixer to $1952 + 488 = 2440$ MHz and output to the antenna. The exciting pulses themselves are controlled by the 1 or 3 Mbps (depending on the device) data bits, effectively amplitude modulating the chip sequences. On reception, such sequences are amplified and downconverted in an image rejection mixer to the 488 MHz IF, amplified again, and applied to the SAW correlator. RF pulses at its output are peak detected and reverted to digital data in a decision circuit.

RFWaves claims high sensitivity (-87 dBm for BER $= 10^{-4}$) and dynamic range as well as low-power consumption. Peak output power is 10 dBm. In addition to the chip sets, the manufacturer offers transceiver modules with a printed loop antenna included on the PC board.

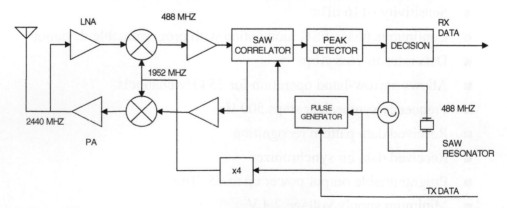

Figure 8-8: RFWaves 2.4 GHZ Spread-Spectrum Transceiver

Micro Linear (www.microlinear.com)

This company produces transceiver chips for the 915, 2400, and 5800 MHz ISM unlicensed bands. Architecture of the three devices, ML2722, ML2724, and ML5800 are similar, based on low IF receivers with image rejection mixers. A fractional-N synthesizer controls the local oscillator and enables accurate FSK modulation in the transmitter. Data rate is 1.5 Mbps and both FHSS and DSSS schemes can be employed. The devices feature automatic alignment of the internal filters and no external filters are needed. Figure 8-9 is a block diagram of the 2.4 GHz ML2724.

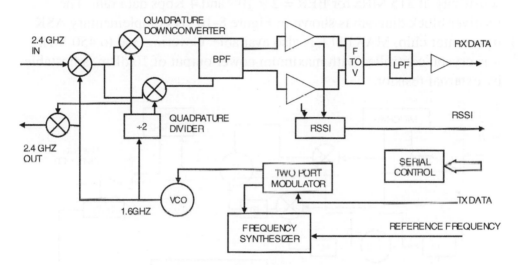

Figure 8-9: Microlinear ML2724 2.4 GHz Transceiver

Maxim (www.maxim.com)

Maxim's MAX1470 is a sensitive ASK receiver with image rejection mixer. It's optimized for 315 MHz operation but can be used from 250 to 500 MHz, including the 433 MHz band. It operates from a supply voltage of 3 to 3.6 volts and operating current is only 5.5 mA. Data slicing and active filter operational amplifiers are integrated, as well as a peak detector for fast tracking of the decision threshold. An internal phase-locked loop and VCO determine the local oscillator frequency at 64 times the reference crystal frequency. The device boasts –115 dBm (average power) sensitivity at 315 MHz for BER = 2×10^{-3} and 4 Kbps data rate. The receiver block diagram is shown in Figure 8-10. A complementary ASK transmitter chip, MAX1472, is also available. Covering 300 to 450 MHz, it is crystal controlled with maximum power output of 10 dBm, adjustable by external resistor.

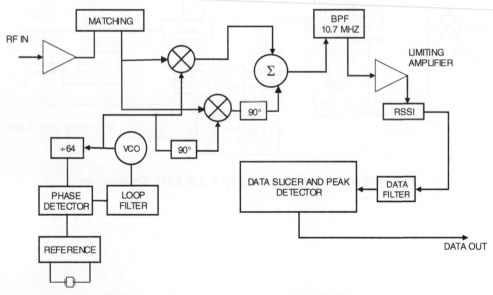

Figure 8-10: Maxim MAX 1470 ASK Receiver

8.3 Large Scale Subsystems

The following three examples are all IF amplifier subsystems. They can be used as the basis for high-performance, short-range communication systems when discrete low noise figure front ends are added, High frequency first IF's may also be required to ensure high image rejection in some of the devices. After detection, on or off the IF subsystem chip, the system designer adds his own baseband filter and data converter specially adopted for the characteristics of the signal he wants to receive.

Other subsystems used in short range radios are receiver front ends—LNA's and mixers, often including the first local oscillator—and frequency synthesizers. For transmitters there are upconverters and power amplifier IC modules.

Philips (www.semiconductors.philips.com)

The NE615 IF amplifier is one example from a whole range of devices that Philips developed for use in analog cellular radios. It consists of an RF mixer which can accept inputs up to 500 MHz., a local oscillator transistor, an IF amplifier and limiter with bandwidth up to 25 MHz, a quadrature FM detector and a muting switch. A received signal strength indicator (RSSI) circuit gives a current output proportional to the signal strength. This output can be used as an ASK detector. The block diagram of the device is shown in Figure 8-11. For more information see Reference 4.

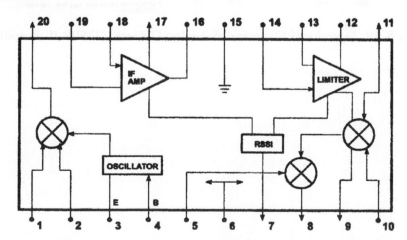

Figure 8-11: Block diagram of Philips NE615 IF Amplifier

Analog Devices (www.analog.com)

Analog Devices' AD608 IF subsystem has similar characteristics to the NE615. Its mixer input accepts up to 500 MHz and its IF bandwidth is suitable for the standard IFs of 455 kHz and 10.7 MHz. Unlike the NE615, the AD608 does not have a built-in quadrature multiplier for FM or FSK modulation. Its RSSI output, which can be used for ASK demodulation, is a voltage source. The AD608 has a power-down terminal to give very low current consumption when the system isn't receiving. Figure 8-12 shows a functional diagram of the AD608.

In comparing the specs of the NE615 and the AD608, it is important to note that the former is characterized for an input frequency of 45 MHz and an IF of 455 kHz, whereas, the latter for inputs of 100 and 240 MHz and an IF of 10.7 MHz. Application notes should be consulted for information about operation under different conditions.

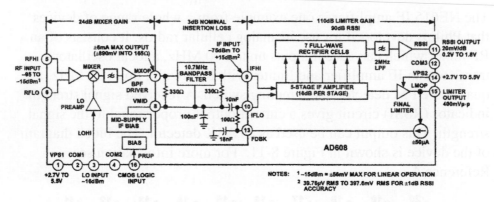

Figure 8-12: Functional Diagram of Analog Devices AD608 IF Amplifier

Maxim (www.maxim.com)

System designers that need special baseband processing or a downconverter and IF strip for a nonstandard transmission frequency may not find suitable the complete transceiver on a chip solutions listed above. Maxim offers two transceiver chips that contain up and down converters plus a high gain IF amplifier and limiter, but without demodulator. The MAX2510 accepts RF or first IF signals at frequencies between 100 and 600 MHz and outputs a limited IF signal at up to 30 MHz. The limiter and RSSI output provide 90 dB of dynamic range. The transmitter has quadrature inputs to suit virtually any type of modulation, 40 dB gain control, and up to 1 dBm output power. The MAX2511 has a similar receiver output, but uses image rejection mixers in both transmitter and receiver and has more restricted frequency ranges of 200 to 440 MHz for RF input and output, and 8 to 13 MHz IF in the receiver and transmitter line. The receiver outputs of both devices can conveniently be connected to an analog-to-digital converter and DSP (digital signal processor) for digital filtering and demodulation.

8.4 Summary

This chapter has presented a sampling of devices that are available to the system designer for developing a short-range radio system. It is far from exhaustive and all of the companies mentioned have other devices which may be more suitable to the designer's need. Most of the companies also have application notes and demonstration boards which can help you decide what device to use and also assist in the design itself. While the availability of a wide range of standard hardware can greatly simplify the design and shorten time-to-market, the system designer must still prepare appropriate interface software and communication protocols, and he is responsible for getting the type approvals in the markets where the product will be sold.

CHAPTER 9

Regulations and Standards

Most short-range radio equipment falls under the category of unlicensed equipment. That means that you don't need a license to use the equipment. However, in most cases, the manufacturer does have to get approval from the relevant authorities to market the devices. The nature of this approval differs from country to country, and equipment which meets approval requirements in one country may not meet them in another. Differences in requirements may be a question of frequency bands, transmitter powers, spurious radiation limits and measurement techniques.

From the point of view of radio regulations, the world is divided into three regions, shown in Figure 9-1. International agreements classify frequency allocations according to these regions, but within each of them, every country is sovereign and makes its own rules controlling radio emissions and the procedures to follow for getting approval. Generally speaking, similar equipment can be sold in the US, Canada, and the other countries of the Americas, whereas this equipment has to be modified for marketing in Europe. We focus in this chapter on the radio regulations in the US and in Europe.

In addition to mandatory requirements that concern the communication qualities of radio equipment, there are also regulations which aim to limit interference from any equipment that radiates electromagnetic waves, intentionally or not. Of major importance is the EMC Directive in the European Community. Radio equipment marketed in the EC must prove the noninterference of its radiation as well as its resistance to misoperation caused by radiation of other devices.

Federal Communications Commission §2.105

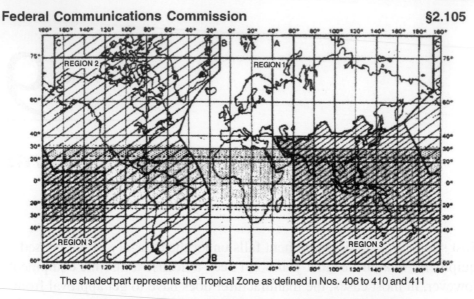

The shaded part represents the Tropical Zone as defined in Nos. 406 to 410 and 411

Figure 1. Chart of the International Regions and Zones as defined in the International Table of Frequency Allocations. [Note: The shaded part represents the Tropical Zone as defined by §2.104b)(4).]

[49 FR 2373, Jan. 19, 1984, as amended at 54 FR 49981, Dec. 4, 1989; 61 FR 15384, Apr. 8, 1996]

Figure 9-1: Division of World into Three Regions for Radio Spectrum Allocation

There are also nonmandatory standards in individual countries that are not a prerequisite for putting a product on sale, but will affect to one degree or another the ability to attract customers. These standards are promulgated by organizations that have the consumer's interest in mind. Examples are the Underwriters Laboratories (UL) in the US and National Approval Council for Security Systems (NACOSS) in the UK.

In the following sections, we review the regulations and standards affecting short-range radio equipment in the US and Europe. Anyone contemplating developing any radio transmitting device should take into account very early in the design the fact that the final product has to meet specific requirements depending on the countries where it will be sold. Usually, the equipment will have to be tested at an approved testing laboratory. Note that regulations undergo changes from time to time, so the relevant official documents should be consulted for authoritative, up-to-date information.

9.1 FCC Regulations

Chapter 47 of the Federal Communications Commission Rules and Regulations sets out the requirements for radio and telecommunication equipment. These requirements are both technical and administrative. Chapter 47 is divided into Parts, each Part dealing with a particular class of equipment. Most of the short-range devices that we are concerned with in this book fall under Part 15, which among other things details the requirements for low-power, unlicensed devices. We will also take a look at other chapters which relate to short-range radio equipment, some of which must be licensed. To start with, we overview Part 2, which has information relating to all of the other Parts.

Part 2—Frequency Allocations and Radio Treaty Matters; General Rules and Regulations

Section 2.1 gives terms and definitions that are referred to throughout the FCC Rules. Some of the important ones for interpreting the requirements for short-range radio are listed in Appendix 9-A of this chapter.

Subpart B—Allocation, Assignment, and Use of Radio Frequencies—forms the bulk of Part 2. Presented in table form are the assignment to various user classes of frequencies in the range of 9 kHz to 400 GHz. The table includes frequency uses for the three world regions, and reference to the relevant Part number for each user class.

Subpart C sets out the nomenclature for defining emission, modulation, and transmission characteristics. It also defines bandwidths. Appendix B of this chapter gives some of the definitions.

Subpart J defines equipment authorization procedures. These may be verification, declaration of conformity, and certification. Verification and declaration of conformity are procedures where the manufacturer, or defined responsible party, makes measurements or takes steps to insure that the equipment complies with the appropriate technical standards. Submittal of samples or test data to the FCC is not required unless specifically requested by the FCC. Receivers generally fall under these categories. In the case of transmitters, a certification procedure must be followed, and the applicant must submit representations and test data to get equipment authorization by the FCC. In practice, what this means is that the equipment has to be tested for compliance with the technical standards by a test facility approved by the FCC. The application for certification is then submitted, along with the test report, by the testing laboratory as a service, or by the applicant himself.

Part 15—Radio Frequency Devices

Part 15 of the FCC regulations "...sets out the regulations under which an intentional, unintentional, or incidental radiator may be operated without an individual license. It also contains the technical specifications, administrative requirements and other conditions relating to the marketing of part 15 devices." (paragraph 15.1 (a)) We review here the paragraphs of Part 15 that are relevant to short-range radio devices as discussed in this book.

Radio receivers are considered unintentional radiators, inasmuch as they produce radio-frequency signals as part of their operation but without the intention to radiate them. The local oscillators (one or more) of superheterodyne receivers produce a frequency in the vicinity of the frequency received (removed from it by the value of the IF) and could interfere with other radio receivers. Superregenerative receivers produce a broadband noisy carrier approximately centered on the received frequency. Tuned radio frequency receivers, including the ASH receiver, do not generate a radio frequency for their operation, but if they contain a microprocessor they are classified as a digital device whose radiation is regulated.

Receiver radiation must meet the following field strength limits, according to paragraph 15.109:

Frequency of Emission (MHz)	Field Strength (microvolts/meter)
30 – 88	100
88 – 216	150
216 – 960	200
Above 960	500

These field strengths are measured at a distance of 3 meters, with the receiver antenna connected.

If the receiver antenna is connected through terminals, the power at the terminals without the antenna may be measured, instead of measuring the radiation from the antenna. The maximum power allowed at the antenna terminals is 2 nW. If this option for checking RF at the antenna is chosen, the limits in paragraph 15.109 must still be checked, with the antenna terminals terminated in the nominal impedance of the antenna.

We now look into the various technical requirements for certifying "intentional radiators" according to Part 15. A common requirement for all part 15 devices is this:

"An intentional radiator shall be designed to ensure that no antenna other than that furnished by the responsible party shall be used with the device. The use of a permanently attached antenna or of an antenna that uses a unique coupling to the intentional radiator shall be considered sufficient to comply with the provisions of this Section. The manufacturer may design the unit so that a broken antenna can be replaced by the user, but the use of a standard antenna jack or electrical connector is prohibited..." (Section 15.203)

Devices that are connected to an AC power line must also comply with limits on conducted radiation up to 30 MHz.

The least restrictive requirements for unlicensed devices regarding their use, the nature of their signals, and frequencies are specified in Section 15.209 reproduced in part below:

Section 15.209 <u>Radiated emission limits, general requirements</u>.

(a) Except as provided elsewhere in this Subpart, the emissions from an intentional radiator shall not exceed the field strength levels specified in the following table:

Frequency (MHz)	Field Strength (microvolts/meter)	Measurement Distance (meters)
0.009 – 0.490	2400/F(kHz)	300
0.490 – 1.705	24000/F(kHz)	30
1.705 – 30.0	30	30
30 – 88	100**	3
88 – 216	150**	3
216 – 960	200**	3
Above 960	500	3

** Except as provided in paragraph (g), fundamental emissions from intentional radiators operating under this Section shall not be located in the frequency bands 54–72 MHz, 76–88 MHz, 174–216 MHz or 470–806 MHz. However, operation within these frequency bands is permitted under other sections of this Part, e.g., Sections 15.231 and 15.241.

(b) In the emission table above, the tighter limit applies at the band edges.

(c) The level of any unwanted emissions from an intentional radiator operating under these general provisions shall not exceed the level of the fundamental emission. For intentional radiators which operate under the provisions of other Sections within this Part and which are required to reduce their unwanted emissions to the limits specified in this table, the limits in this table are based on the frequency of the unwanted emission and not the fundamental frequency. However, the level of any unwanted emissions shall not exceed the level of the fundamental frequency.

(d) The emission limits shown in the above table are based on measurements employing a CISPR quasi-peak detector except for the frequency bands 9-90 kHz, 110–490 kHz and above 1000 MHz. Radiated emission limits in these three bands are based on measurements employing an average detector.

....

(g) Perimeter protection systems may operate in the 54–72 MHz and 76–88 MHz bands under the provisions of this section. The use of such perimeter protection systems is limited to industrial, business and commercial applications.

The field strengths for frequencies between 216 and 960 MHz, which are our primary interest, are quite low and suffice for reliable communication only over a few meters, which may well be sufficient for some classes of short-range devices—a wireless mouse or keyboard, for example. The 200 µV/m limit at 3 meters translates to radiated power of only 12 nW.

Subsection (d) refers to a CISPR quasi-peak detector. This instrument has specified attack times, decay times, and 6 dB bandwidths as well as overload factors which are shown as a function of frequency range in the following table (Reference 19, Draft P802.15.4/D18 Section F):

	9–150 kHz	*0.15–30 MHz*	*30–1000 MHz*
6 Db BW (kHz)	0.2	9	120
Attack time (ms)	45	1	1
Decay time (ms)	500	160	550
Pre-Detector overload factor (dB)	24	30	43.5

On specified frequency bands, higher power is allowed, although with restrictions as to use. Below are highlights of the regulations, with comments, for short-range communicating devices, presented according to Part 15 Section.

15.229 Operation within the band 40.66-40.70

Operation on this band is possible with a field strength limitation of 1000 µV/m at 3 meters, as compared to 100 µV/m at 3 meters according to Section 15.209. Oscillator stability must be within 100 ppm. The equivalent radiated power is 300 nW. Potential users should note that the background noise at this low VHF frequency is comparatively high, compared to the more popular UHF bands. Note that there are no restrictions to the type of operation according to this section, so, for example, continuous voice transmissions could be used, which are prohibited under Section 15.231 (see below).

15.231 Periodic operation in the band 40.66 – 40.70 and above 70 MHz

Most of the short-range wireless links for security and medical systems and keyless entry are certified to the requirements in this section. Considerably more field strength is allowed than in 15.209, and there is a wide choice of frequencies. However, the transmission duty cycle is restricted and voice and video transmissions are not allowed. This is not a serious impediment for security systems, except in regard to sending periodic supervision or status signals, which are limited to two seconds per hour. The provisions of this section are reproduced below, as updated in the release of 22 July 2003. There were two important changes in this release—transmission of data is permitted along with the control signal, and the frequency of polling transmissions is not restricted as long as their durations do not exceed two seconds per hour per transmitter.

<u>Section 15.231 Periodic operation in the band 40.66 – 40.70 MHz and above 70 MHz.</u>

(a) The provisions of this section are restricted to periodic operation within the band 40.66–40.70 MHz and above 70 MHz. Except as shown in paragraph (e) of this Section, the intentional radiator is restricted to the transmission of a control signal such as those used with alarm systems, door openers, remote switches, and so forth. Continuous transmissions, voice, video and the radio control of toys are not permitted. Data is permitted to be sent with a control signal. The following conditions shall be met to comply with the provisions for this periodic operation:

(1) A manually operated transmitter shall employ a switch that will automatically deactivate the transmitter within not more than 5 seconds of being released.

(2) A transmitter activated automatically shall cease transmission within 5 seconds after activation.

(3) Periodic transmissions at regular predetermined intervals are not permitted. However, polling or supervision transmissions, including data, to determine system integrity of transmitters used in security or safety applications are allowed if the total duration of transmissions does not exceed more than two seconds per hour for each transmitter. There is no limit on the number of individual transmissions, provided the total transmission time does not exceed two seconds per hour.

(4) Intentional radiators which are employed for radio control purposes during emergencies involving fire, security, and safety of life, when activated to signal an alarm, may operate during the pendency of the alarm condition.

(b) In addition to the provisions of Section 15.205, the field strength of emissions from intentional radiators operated under this Section shall not exceed the following:

Fundamental Frequency (MHz)	Field Strength of Fundamental (microvolts/meter)	Field Strength of Spurious Emissions (microvolts/meter)
40.66 – 40.70	2,250	225
70 – 130	1,250	125
130 – 174	1,250 to 3,750**	125 to 375**
174 – 260	3,750	75
260 – 470	3,750 to 12,500**	375 to 1,250**
Above 470	12,500	1,250

** linear interpolations

[Where F is the frequency in MHz, the formulas for calculating the maximum permitted fundamental field strengths are as follows: for the band 130–174 MHz, μV/m at 3 meters = 56.81818(F) – 6136.3636; for the band 260–470 MHz, μV/m at 3 meters = 41.6667(F) – 7083.3333. The maximum permitted unwanted emission level is 20 dB below the maximum permitted fundamental level.]

(1) The above field strength limits are specified at a distance of 3 meters. The tighter limits apply at the band edges.

(2) Intentional radiators operating under the provisions of this Section shall demonstrate compliance with the limits on the field strength of emissions, as shown in the above table, based on the average value of the measured emissions. As an alternative, compliance with the limits in the above table may be based on the use of measurement instrumentation with a CISPR quasi-peak detector. The specific method of measurement employed shall be specified in the application for equipment authorization. If average emission measurements are employed, the provisions in Section 15.35 for averaging pulsed emissions and for limiting peak emissions apply. Further, compliance with the provisions of Section 15.205 shall be demonstrated using the measurement instrumentation specified in that section.

(3) The limits on the field strength of the spurious emissions in the above table are based on the fundamental frequency of the intentional radiator. Spurious emissions shall be attenuated to the average (or, alternatively, CISPR quasi-peak) limits shown in this table or to the general limits shown in Section 15.209, whichever limit permits a higher field strength.

(c) The bandwidth of the emission shall be no wider than 0.25% of the center frequency for devices operating above 70 MHz and below 900 MHz. For devices operating above 900 MHz, the emission shall be no wider than 0.5% of the center frequency. Bandwidth is determined at the points 20 dB down from the modulated carrier.

(d) For devices operating within the frequency band 40.66 – 40.70 MHz, the bandwidth of the emission shall be confined within the band edges and the frequency tolerance of the carrier shall be ± 0.01%. This frequency tolerance shall be maintained for a temperature variation of –20 degrees to +50 degrees C at normal supply voltage, and for a variation in the primary supply voltage from 85% to 115% of the rated supply voltage at a temperature of 20 degrees C. For battery operated equipment, the equipment tests shall be performed using a new battery.

(e) Intentional radiators may operate at a periodic rate exceeding that specified in paragraph (a) and may be employed for any type of operation, including operation prohibited in paragraph (a), provided the intentional radiator complies with the provisions of paragraphs (b) through (d) of this Section, except the field strength table in paragraph (b) is replaced by the following:

Fundamental Frequency (MHz)	Field Strength of Fundamental (microvolts/meter)	Field Strength of Spurious Emissions (microvolts/meter)
40.66 – 40.70	1,000	100
70 – 130	500	50
130 – 174	500 to 1,500**	50 to 150**
174 – 260	1,500	150
260 – 470	1,500 to 5,000**	150 to 500**
Above 470	5,000	500

** linear interpolations

[Where F is the frequency in MHz, the formulas for calculating the maximum permitted fundamental field strengths are as follows: for the band 130–174 MHz, μV/m at 3 meters = 22.72727(F) – 2454.545; for the band 260–470 MHz, μV/m at 3 meters = 16.6667(F) – 2833.3333. The maximum permitted unwanted emission level is 20 dB below the maximum permitted fundamental level.]

In addition, devices operated under the provisions of this paragraph shall be provided with a means for automatically limiting operation so that the duration of each transmission shall not be greater than one second and the silent period between transmissions shall be at least 30 times the duration of the transmission but in no case less than 10 seconds.

Comments on Section 15.231

a) Section 15.205 referred to in the present section contains a table of restricted frequency bands where fundamental transmissions are not allowed and where spurious limits may be lower than those presented above. This table is reproduced in Appendix 9-C.

b) The field strength limits in this section are based on average measurements. Thus, when ASK is used, peak field strength may exceed average field strength by a factor that is the inverse of the duty cycle of the modulation. However, Section 15.35 limits the peak-to-average ratio to 20 dB. The method of determining the average field strength is stated in Section 15.35 as follows:

"…when the radiated emission limits are expressed in terms of the average value of the emission, and pulsed operation is employed, the measurement field strength shall be determined by averaging over one complete pulse train, including blanking intervals, as long as the pulse train does not exceed 0.1 seconds. As an alternative (provided the transmitter operates for longer than 0.1 seconds) or in cases where the pulse train exceeds 0.1 seconds, the measured field strength shall be determined from the average absolute voltage during a 0.1 second interval during which the field strength is at its maximum value. The exact method of calculating the average field strength shall be submitted with any application for certification or shall be retained in the measurement data file for equipment subject to notification or verification."

15.235 Operation within the band 49.82 – 49.90 MHz

(a) The field strength of any emission within this band shall not exceed 10,000 microvolts/meter at 3 meters. The emission limit in this paragraph is based on measurement instrumentation employing an average detector. The provisions in Section 15.35 for limiting peak emissions apply.

(b) The field strength of any emissions appearing between the band edges and up to 10 kHz above and below the band edges shall be attenuated at least 26 dB below the level of the unmodulated carrier or to the general limits in Section 15.209, whichever permits the higher emission levels. The field strength of any emissions removed by more than 10 kHz from the band edges shall not exceed the general radiated emission limits in Section 15.209. All signals exceeding 20 microvolts/meter at 3 meters shall be reported in the application for certification.

This low VHF band has an advantage over the operation permitted under Section 15.231 because it can be used for continuous transmissions for either analog or digital communication. Permitted average field strength at 3 meters is 10,000 μV/m, which is capable of attaining a range of several hundred meters if an effective receiving antenna is used. The allowed bandwidth is only 80 kHz so transmitters and receivers must have crystal-controlled oscillators.

15.237 Operation in the bands 72.0 – 73 MHz, 74.6 – 74.8 MHz and 75.2 – 76 MHz

This section is restricted to auditory assistance devices, defined as

"An intentional radiator used to provide auditory assistance to a handicapped person or persons. Such a device may be used for auricular training in an educational institution, for auditory assistance at places of public gatherings, such as a church, theater, or auditorium, and for auditory assistance to handicapped individuals, only, in other locations." (Section 15.3 (a))

Maximum allowed bandwidth is 200 kHz and the limit on average field strength is 80 mV/m at 3 meters.

15.241 Operation in the band 174 – 216 MHz

Operation under this Section is restricted to biomedical telemetry devices, defined as

"An intentional radiator used to transmit measurements of either human or animal biomedical phenomena to a receiver." (Section 15.3 (b))

Maximum bandwidth is 200 kHz and the average field strength limit is 1500 μV/m at 3 meters.

15.242 Operation in the bands 174 – 216 MHz and 470 – 668 MHz

This section also is dedicated to biomedical telemetry devices but they must be employed solely on the premises of health care facilities, which includes hospitals and clinics. Signals of bandwidths of 200 kHz and up to 200 mV/m at 3 meters are allowed. This section defines in detail the requirements for avoiding interference with television stations and radio astronomy observatories.

15.247 Operation within the bands 902 – 928 MHz, 2400 – 2483.5 MHz, and 5725 – 5850 MHz

This section defines the requirements for devices using spread spectrum and other wideband modulation in the ISM bands. Devices designed to comply with the requirements of this section have the potential for highest performance, in respect to range and interference immunity, of all unlicensed equipment. We present here a summary of the major points of this section. For details, refer to the source.

Both frequency hopping and what is called digital modulation is allowed, as well as a combination of the two methods. Digital modulation is defined as (Section 15.403(b)): "The process by which the characteristics of a carrier wave are varied among a set of predetermined discrete values in accordance with a digital modulating function..."

Previous to the revision dated August 2002, this section referred to direct sequence spread-spectrum modulation but that term was dropped so as not to restrict the introduction of new types of wideband modulation methods. In particular, the modulation systems introduced to increase data rates in WLANs according to IEEE specifications 802.11*b*, *g*, and *a* do not fit the classical definition of direct sequence spread-spectrum but were deemed to be suitable for inclusion in this FCC section (see Chapter 11). The technical requirements of this section differ according to the type of spectrum-spreading (frequency-hopping or digital) and the particular frequency band.

1) Frequency-hopping spread-spectrum

Minimum hopping channel separation:	25 kHz, but not less than the channel bandwidth at 20 dB down.
Minimum number of pseudorandomly ordered hopping frequencies:	a) 900 MHz band—50 for hopping channel BW less than 250 kHz and 25 for BW greater than 250 kHz. Max. BW 500 kHz at 20 dB down. b) 2.4 GHz band—15 nonoverlapping channels. Intelligent hopping (to avoid interference) may be used when fewer than 75 channels are employed. c) 5.7 GHz band—75 hopping frequencies. Max BW 1 MHz at 20 dB down.
Occupancy time on channel frequency:	a) 900 MHz band—Max. 0.4 sec. within 20 sec. period for BW less than 250 kHz, and 0.4 sec. within 10 sec. for BW greater than 250 kHz. b) 2.4 GHz band—Max. 0.4 sec. within 0.4 sec times number of hopping channels used. c) 5.7 GHz band—Max. 0.4 sec. within 30 sec. period.
Maximum peak output power:	a) 900 MHz band—1 watt when using at least 50 hopping channels; 0.25 watt with less than 50 hopping channels. b) 2.4 GHz band—1 watt when using at least 75 hopping channels, otherwise 0.125 watt. c) 5 GHz band—1 watt NOTE: For all bands, except for 2.4 and 5 GHz band fixed point-to-point applications, when using directional antennas the maximum peak power limit is reduced by the amount that the antenna gain exceeds 6 dBi.

2) Digital Modulation

Minimum bandwidth:	500 kHz at 6 dB down
Maximum peak output power:	1 watt
Peak power spectral density to the antenna:	8 dBm in any 3-kHz band

15.249 Operation within the bands 902 – 928 MHz, 2400 – 2483.5 MHz, 5725 – 5875 MHz, and 24.0 – 24.25 GHz

This subsection regulates operation in upper UHF and microwave bands without specifying the type of use or modulation constraints. Relevant provisions relating to the three lower bands are reproduced below.

Section 15.249 Operation within the bands 902 – 928 MHz, 2400 – 2483.5 MHz, 5725 – 5875 MHz, and 24.0 – 24.25 GHz.

(a) Except as provided in paragraph (b) of this section, the field strength of emissions from intentional radiators operated within these frequency bands shall comply with the following:

Fundamental Frequency	Field Strength of Fundamental (millivolts/meter)	Field Strength of Harmonics (microvolts/meter)
902 – 928 MHz	50	500
2400 – 2483.5 MHz	50	500
5725 – 5875 MHz	50	500
24.0 – 24.25 GHz	250	2500

(b) [*Relates to fixed point-to-point operation on the 24.05 to 24.25 GHz band.*]

(c) Field strength limits are specified at a distance of 3 meters.

(d) Emissions radiated outside of the specified frequency bands, except for harmonics, shall be attenuated by at least 50 dB below the level of the fundamental or to the general radiated emission limits in Section 15.209, whichever is the lesser attenuation.

(e) As shown in Section 15.35(b), for frequencies above 1000 MHz, the above field strength limits in paragraphs (a) and (b) of this section are based on average limits. However, the peak field strength of any emission shall not exceed the maximum permitted average limits specified above by more than 20 dB under any condition of modulation.

If contemplating the use of this section, note the following:

1) Harmonic radiation must be reduced 40 dB compared to the maximum allowed fundamental, and not 20 dB as on the lower UHF bands.

2) The field strength on the 900-MHz band is determined by a quasi-peak or peak detector, and is not an average value as in Section 15.231 referenced above. This reduces the possible advantage of using ASK with low duty cycle.

3) The detector bandwidth for 902 – 928 MHz band measurements is 120 kHz. Since field strength is measured over this bandwidth, signals with wider bandwidths can have higher total field strength, or equivalent power, depending on the degree that their bandwidth is greater than 120 kHz and on the shape of the signal's spectrum.

Subpart E—Unlicensed National Information Infrastructure Devices (U-NII)

A special category was created in FCC Part 15 for "Intentional radiators operating in the frequency bands 5.15 – 5.35 GHz and 5.725 – 5.825 GHz that use wideband digital modulation techniques and provide a wide array of high data rate mobile and fixed communications for individuals, businesses, and institutions." Our interest in U-NII primarily relates to their use in WLANs, specifically in connection with specification IEEE 802.11a (see Chapter 11). While the wide growth of WLAN is occurring with equipment operating in the 2.4 to 2.4835 GHz ISM band, problems of interference with other types of wireless services and restricted bandwidth are making the 5 GHz U-NII an attractive alternative for continued development. With a more restricted definition of allowed applications and a total bandwidth of 300 MHz as compared to the 83.5 MHz available on 2.4 GHz, U-NII offers an outlet for high performance wireless network connectivity that won't approach saturation for many years to come.

The principle characteristics of U-NII devices as put forth in Section 15.407 of FCC Part 15 are summarized in the following table:

	5.15–5.25 GHz	*5.25–5.35 GHz*	*5.725–5.825 GHz*
Maximum power to antenna*	50 mW	250 mW	1 Watt
Antenna gain restriction	Maximum allowed power is reduced by the amount of antenna gain over 6 dBi	Maximum allowed power is reduced by the amount of antenna gain over 6 dBi	Antenna gain up to 23 dBi with no power reduction is allowed for fixed point-to-point systems—otherwise allowed power is reduced for antenna gain over 6 dBi
Special restrictions	Indoor use only; must use integral transmitter antenna		

*NOTE: Stated maximum powers are specified for 26 dB bandwidth of 20 MHz or greater. Power limits are reduced continuously for lower bandwidths.

Subpart F—Ultra-Wideband (UWB) Operation

As of the FCC Part 15 revision of August 2002, a new category of short-range communication is included in the standard—ultra-wideband. This category differs from all the other intentional radiation sections by being defined in a much wider spectrum allocation in relation to the median frequency of the band. Chapter 11 gives a description of ultra-wideband transmissions.

The bandwidth of UWB is defined by the FCC as the difference between the two frequencies on both sides of the frequency of maximum radiation at which the radiated emission is 10 dB down. The bandwidth is based on the whole transmission system, including the antenna. If the upper and lower 10 dB down frequencies are f_H and f_L respectively, then the fractional bandwidth equals $2(f_H - f_L)/(f_H + f_L)$. The center frequency is defined as $f_C = (f_H + f_L)/2$. A transmitter that falls in the category of UWB has a fractional bandwidth equal to or greater than 0.2 or a UWB bandwidth equal to or greater than 500 MHz, regardless of fractional bandwidth.*

* In a "Memorandum Opinion and Order and Further Notice of Proposed Rulemaking on Ultra-wideband," dated 13 March 2003, the FCC proposed to eliminate this definition and to permit the operation of any transmission system, regardless of its bandwidth, as long as it complies with the standards for UWB operation of Part 15, Subpart F.

FCC Part 15 Subpart F specifies several categories of UWB systems, some of which must be licensed and are intended for specific users. For example, UWB imaging systems, used for through the wall imaging, ground penetration radar, medical imaging (detects location or movement of objects within a living body), and surveillance, may be marketed only to law enforcement, rescue, medical, or certain industrial or commercial entities who are solely eligible to operate the equipment. Our concern is with two more general categories—indoor and handheld UWB systems.

Both indoor and handheld devices must have UWB bandwidths contained in the frequency range of 3,100 to 10,600 MHz. Radiation at or below 960 MHz must not exceed the levels of Section 15.209 (see above). Above 960 MHz, emission limits for the two categories are according to the following table:

	Indoor	*Handheld*
Frequency MHz	*Average EIRP dBm*	*Average EIRP dBm*
960–1610	−75.3	−75.3
1610–1990	−53.3	63.3
1990–3100	−51.3	−61.3
3100–10600	−41.3	−41.3
Above 10600	−51.3	−61.3

The resolution bandwidth of the measuring device is 1 MHz.

In order to specifically protect GPS (Global Positioning System) receivers from UWB interference, radiation around the two GPS channels is checked with narrow-resolution bandwidth as follows:

Frequency MHz	*Average EIRP dBm*
1164–1240	−85.3
1559–1610	−85.3

This table applies to both indoor and handheld devices. Resolution bandwidth is no less than 1 kHz.

Peak emissions within a 50 MHz bandwidth around the frequency of highest radiation are limited to 0 dBm EIRP.

Part 74 Experimental Radio, Auxiliary, Special Broadcast and Other Program Distributional Services

Subpart H regulates low-power auxiliary stations, defined as follows:

Low-power auxiliary station. An auxiliary station authorized and operated pursuant to the provisions set forth in this subpart. Devices authorized as low-power auxiliary stations are intended to transmit over distances of approximately 100 meters for uses such as wireless microphones, cue and control communications, and synchronization of TV camera signals.

Frequency assignments listed in Section 74.802 cover frequency bands from 26.1 MHz up to 952 MHz. Popular bands for wireless microphones are 174 MHz to 216 MHz in the VHF range and 774 MHz to 806 MHz, UHF. The regulations specify requirements to avoid interference to TV channels.

Users of devices covered in this Part must obtain a license from the FCC. Eligible for licenses are radio and television station licensees, cable television station operators that produce live programs, and television broadcasting and motion picture producers.

Maximum power output allowed for VHF transmitters is 50 mW and for UHF transmitters 250 mW. Up to 200 kHz bandwidth and ±75 kHz deviation is allowed.

Part 95 Personal Radio Services

Some of the provisions of Part 95 that apply to unlicensed short-range communication are given in the following paragraphs.

Subpart B describes the general provisions of the Family Radio Service (FRS). This is a license-free service intended for two-way communication only. The basic technical requirements are:

- Frequency: 14 channels in the ranges of 462.5625 to 462.7125 MHz and 467.5625 to 467.7125 MHz.

- Modulation: FM only.

- Bandwidth: 12.5 kHz.

- Power: .5 Watt ERP.

Subpart G gives the provisions for the Low Power Radio Service (LPRS), which is part of the Citizens Band Radio Service. An operator of a transmitter meeting the restrictions of this subpart is not required to be individually licensed by the FCC.

LPRS stations may transmit voice, data, or tracking signals as permitted in section 95.1009. Two-way voice communications are prohibited. These are some of the uses permitted:

1) Auditory assistance communications including assistive listening devices, audio description for the blind, and simultaneous language translation. Beneficiaries of this use are

a) Persons with disabilities

b) Persons who require language translations

c) Persons who may otherwise benefit from auditory assistance communications in educational settings

2) Health-care-related communications for the ill

3) Law enforcement tracking signals (for homing or interrogation) including the tracking of persons or stolen goods under authority or agreement with a law enforcement agency (federal, state, or local) having jurisdiction in the area where the transmitters are placed.

LPRS equipment can operate on designated frequency channels between 216 MHz and 217 MHz. Channels are designated for bandwidths of 5 kHz, 25 kHz and 50 kHz. Maximum authorized transmitter output power is 100 mW ERP.

Part 95 regulates low power transmissions for medical services, in two categories. Subpart H defines the Wireless Medical Telemetry Service (WMTS). Subpart I relates to Medical Implant Communications (MICS).

The frequency bands 608 to 614 MHz, 1395 to 1400 MHz and 1427 to 1429.5 MHz are allocated on a nonexclusive basis for telemetry transmissions related to medical services. Operation of equipment requires prior coordination with an FCC frequency coordinator for the Wireless Medical Telemetry Service. Any appropriate emission types are allowed, except for video and voice. Waveforms such as electrocardiograms are not considered video. The maximum field strength limit in the 608 to 614 MHz band, measured with a CISPR quasi-peak detector at 3 meters, is 200 mV/m. Over 1395 to 1400 MHz and 1427 to 1429.5 MHz, field strength is limited to 740 mV/m at 3 meters, using an averaging detector and a 1 MHz measurement bandwidth. Frequency stability must be such that emission is maintained within the band of operation under all of the manufacturer's specified conditions.

Medical Implant Communications are specified for the operation of medical implant devices as well as the programmer/control transmitters associated with them. An individual FCC license is not required. These are the basic technical requirements:

■ Frequency band: 402 to 405 MHz

■ Maximum emission bandwidth: 300 kHz total. Bandwidth is measured between points 20 dB down from the maximum carrier level.

■ Power output: 25 microwatt EIRP. Radiation from implantation transmitters must be tested in a medium that simulates human tissues.

■ Frequency stability: ±100 ppm

■ Emission type: any except voice

9.2 Test Method for Part 15

The method for measuring field strength for radio control and security alarm devices and receivers is described in FCC document MP-1. The equipment arrangement is shown in Figure 9-2. Measurements are made in an FCC-approved open field test site. The equipment under test (EUT) is placed on a wooden turntable, one meter high, located three meters from the test antenna. Measurements are taken on a field strength measuring receiver or spectrum analyzer. Test instrument bandwidth must be at least 100 kHz for signals below 1 GHz and 1 MHz for signals above 1 GHz. When peak readings are taken, the average value for digital signals can be calculated from the baseband pulse diagram. Reading are taken at the fundamental frequency and all spurious radiation frequencies.

Field strength readings are taken at all practical equipment orientations by rotating the turntable. The height of the test antenna is varied from between one to four meters and readings are taken for both vertical and horizontal antenna orientations. At each emission frequency, the maximum reading is recorded for all possibilities of equipment orientation, test antenna height and polarization.

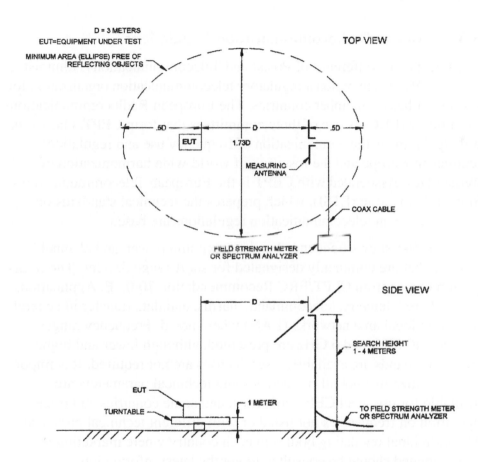

Figure 9-2: Equipment Arrangement for FCC Part 15 Tests

Note the importance, in developing a short-range radio transmitter, of obtaining as far as possible omnidirectional radiation characteristics of the built-in antenna. Since the power of the device is limited at its position of maximum radiation, a transmitter with unintentional directional characteristics will give less than maximum range when the orientation of the transmitter relative to the receiver is uncontrollable, as in most applications.

9.3 European Radiocommunication Regulations

The European Conference of Postal and Telecommunications Administrations (CEPT) is the regional regulatory telecommunication organization for Europe. It has 43 member countries. The European Radiocommunications Committee (ERC) is one of three committees that form CEPT. One of its principal aims is the harmonization of frequency use and regulation regimes in Europe and the fostering of world-wide harmonization of frequencies. Associated with CEPT is the European Telecommunications Standards Institute (ETSI), which prepares the technical standards on which European telecommunication regulations are based.

Table 9-1 shows frequency bands, maximum power, and channel spacing that are commonly designated for short-range devices. The information comes from CEPT/ERC Recommendation 70-03 E. Applications covered are telemetry, telecommand, alarms, and data transfer in general, including local area networks (LAN) where noted. Frequency ranges between 40 MHz and 6 GHz are presented, although lower and higher frequency bands are available. User licenses are not required. It is important to realize that not all frequencies and technical parameters are available for use in all CEPT countries, and some countries authorize operation on frequencies not listed or with different technical parameters. The individual regulating bodies in each country where marketing is contemplated should be consulted to get the latest information.

Table 9-1

Frequency Band (MHz)	Maximum Power	Channel Spacing	Duty Cycle (%)
40.660–40.700	10 mW e.r.p.	None (whole band)	No restriction
138.2–138.45	10 mW e.r.p.	None	1
433.050–434.790	10 mW e.r.p.	None	10
868.0–868.6	25 mW e.r.p.	25 kHz or wide band	1
868.6–868.7 (alarms exclusively)	10 mW e.r.p.	25 kHz or high-speed data	0.1
868.7–869.2	25 mW e.r.p.	25 kHz or wide band	0.1
869.2–869.25 (*social alarms exclusively)	10 mW e.r.p.	25 kHz	0.1
869.25–869.30	10 mW e.r.p.	25 kHz	0.1
869.3–869.4	10 mW	25 kHz	No restriction
869.4–869.65	500 mW e.r.p.	25 kHz or whole channel for high-speed data	10
869.65–869.70 (alarms exclusively)	25 mW e.r.p.	25 kHz	10
869.7–870.0	5 mW e.r.p.	None	Up to 100
2400–2483.5	10 mW e.i.r.p.	None	No restriction
2400–2483.5 (for LAN and Spread Spectrum modulation)	100 mW e.i.r.p.	None	No restriction
5725–5875	25 mW e.i.r.p.	None	No restriction
5150–5350 HIPERLAN (indoor)	200 mW	None	No restriction
5470–5725 HIPERLAN	1 W	None	No restriction

*Social alarm systems are intended to assist elderly or disabled people living at home when they are in a distress situation.

We review below the ETSI specifications that are relevant to short-range radio communications. Each country in the European Community may or may not base their regulations wholly on these specifications, so it's necessary to check with the regulating authorities in each country where the product will be sold.

EN 300 220-1 Technical characteristics and test methods for radio equipment to be used in the 25-MHz to 1000-MHz frequency range with power levels ranging up to 500 mW.

The document details test methods for a range of short-range devices and applications within the frequency range and power levels of its title. Parameters to be measured are:

- Output power
- Response to modulation (AM or FM)
- Adjacent channel power (narrow-band emissions)
- Modulation bandwidth (wide-band emissions)
- Spurious emissions
- Frequency stability over supply voltage limits
- Duty cycle

The method used to measure transmitter emissions depends on whether the device has an antenna connector or an integral antenna that cannot be disconnected. In the former case, radiation characteristics can be measured directly across a dummy load on the RF output connector. We describe here the principle of the method used to determine power output, both fundamental and spurious, by measuring effective radiated power (refer to the standard for authoritative details).

The test site is arranged as shown in Figure 9-3. The distance between the equipment under test and the test antenna is approximately 3 meters. It is not critical. Reflecting objects around the site must be far enough away so that they don't influence the measurements. The effective radiated power at the fundamental frequency and harmonics and spurious frequencies is determined in two stages as follows.

The equipment under test is powered and the receiving instrument, receiver or spectrum analyzer, indicates the level received by the test antenna. A CW carrier from the transmitter is preferred but if data modulation is used, the bandwidth of the measuring receiver is set to 120 kHz for frequencies under 1 GHz and 1 MHz for frequencies over 1 GHz (spurious radiation). A peak reading (CW) or quasi-peak (modulated) is used. The equipment is rotated and the test antenna height is varied between one and four meters to find the maximum level, which is noted.

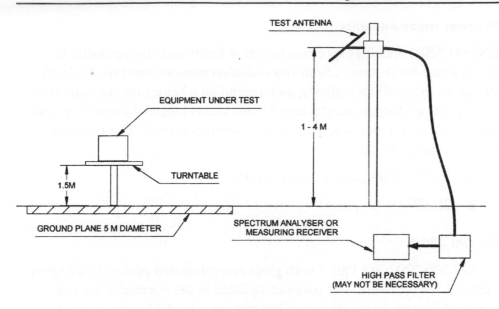

Figure 9-3: EN300 220 Test Site

In the second stage, the equipment under test is replaced by a half-wave dipole antenna connected to a signal generator. The height of the test antenna is adjusted for maximum output. Then the generator output is adjusted so that the receiver indicates the same maximum level as obtained in the first stage. The power level of the generator is recorded.

The test is carried out for vertical and horizontal polarization. The higher of the recorded generator power outputs is the effective radiated power of the equipment under test.

For fundamental frequency power measurements, the effective radiated power shall not exceed the maximum value as shown in Table 9-1 or otherwise permitted by the relevant authority. For spurious outputs the limits are as shown in Table 9-2.

Table 9-2

State	47 MHz to 74 MHz 87.5 MHz to 118 MHz 174 MHz to 230 MHz 470 MHz to 862 MHz	Other frequencies below 1000 MHz	Frequencies above1000 MHz
Operating	4 nW	250 nW	1 µW
Standby	2 nW	2 nW	20 nW

Receiver measurements

EN 300 220-1 specifies spurious radiation limits and susceptibility to interference for receivers. Spurious radiation measurement methods are similar to those of transmitters, and depend on whether the measurement is through conduction, used when the antenna is attached through a connector, or by radiation when an integral antenna is used. The limits of spurious radiation are:

- 2 nW for frequencies up to 1 GHz
- 20 nW for frequencies above 1 GHz

EN 300 220-2

EN 300 220 has a Part 2 with gives recommended performance specifications for receivers. The parameters listed in the document are not intended for regulatory purposes, but can serve as guidelines to short-range radio designers both as regards the specification limits themselves and the way they are defined. The parameters covered are:

- Maximum usable sensitivity (conducted)
- Average usable sensitivity (field strength)
- Co-channel reflection
- Adjacent channel selectivity
- Spurious response rejection
- Intermodulation response rejection
- Blocking or desensitization

EN 300 440 Technical characteristics and test methods for radio equipment to be used in the 1 GHz to 40 GHz frequency range

This is a generic standard for short-range radio equipment transmitting on frequencies above 1 GHz. It covers all types of modulation, with or without speech. Peak power levels up to 5 watts e.i.r.p. are covered. Remember that the e.i.r.p. is the output power of the transmitter times the gain of the antenna.

The method of measuring peak power depends on whether the signal bandwidth is less than or more than 1 MHz. When the bandwidth is less than 1 MHz, peak power can be found directly from the spectrum analyzer. At higher bandwidths, peak power is found by making an average power measurement with a power meter and calculating peak power using the observed or declared duty cycle.

Power measurements may be made either by conduction through the antenna connector or, if there is no connector, by radiation and test and substitution antennas, as in EN 300 220.

EN 300 440 gives basic requirements for frequency-hopping spread-spectrum (FHSS) modulation. There must be at least 20 hopping frequencies and maximum dwell time on one frequency channel is 0.4 seconds.

ETS 300 328 Technical characteristics and test conditions for data transmission equipment operating in the 2.4-GHz ISM band and using spread-spectrum modulation techniques

This specification covers wide-band data transmission equipment having the following technical characteristics:

- Aggregate bit rates in excess of 250 kbits/s

- Operation in the 2.4 to 2,483.5 GHz ISM band

- Effective radiated power of up to 100 mW

- Power density of up to 100 mW per 100 kHz for frequency hopping modulation

- Power density of up to 10 mW per 1 MHz for other forms of spread-spectrum modulation

The standard covers two types of spread-spectrum modulation techniques:

a) Frequency-hopping spread spectrum (FHSS) with the following characteristics:

- At least 20 hopping channels

- 1 MHz maximum 20 dB bandwidth of single hop channel

- 0.4 second maximum dwell time on a channel

■ Each channel occupied at least once during period not exceeding 4 times dwell time times number of channels

b) Direct-sequence spread spectrum (DSSS), defined here as all other forms of spread spectrum which are not FHSS according to the definition above.

The signal characteristics defined in the standard are these:

■ Maximum power: 100 mW e.i.r.p

■ Peak power density: 100 mW/100 kHz e.i.r.p for FHSS and 10 mW/1 MHz e.i.r.p. for DSSS

■ Frequency range: 2.4 to 2.483 GHz

■ Spurious emission limits are prescribed for transmitter and receiver

The standard details the measurement methods for each parameter.

9.4 The European Union Electromagnetic Compatibility Requirements

Electrical and electronic equipment destined for marketing in the member countries of the European Union must comply with the Radio and Telecommunications Terminal Equipment (R&TTE) Directive, 1999/5/EEC. The R&TTE Directive, whose authority commenced on April 8, 2000, signified a new approach to the legal aspects of marketing radio and telecommunication equipment in Europe. Because of its importance to any company that intends to sell short-range devices within the European community, we review below in some detail the major provisions of the Directive that particularly affect this category of equipment.

The aim of the Directive is stated in Article 1 as the establishment of "a regulatory framework for the placing on the market, free movement and putting into service in the Community of radio equipment and telecommunications terminal equipment." All technical specifications and characteristics of equipment falling under the scope of the Directive revolve around the following "essential requirements" and "addition," quoted below from Article 3.

1. The following essential requirements are applicable to all apparatus:

(a) the protection of the health and the safety of the user and any other person, including the objectives with respect to safety requirements contained in Directive 73/23/EEC, but with no voltage limit applying;

(b) the protection requirements with respect to electromagnetic compatibility contained in Directive 89/336/EEC.

2. In addition, radio equipment shall be so constructed that it effectively uses the spectrum allocated to terrestrial/space radio communication and orbital resources so as to avoid harmful interference.

The EMC Directive referred to in 1(b) itself specifies two essential requirements:

a) The device cannot interfere with radio or telecommunications equipment operation;

b) the device itself must be immune from electromagnetic disturbance due to sources such as RF transmitters and other equipment.

The requirements of the Directive are considered to be met if the equipment complies with harmonized R&TTE standards, when they exist, published periodically in the *Official Journal of the European Union*. As opposed to the approval regime that was in force previously, manufacturers do not apply to the communication authorities of the individual EU countries for authorization but they themselves bear full responsibility for meeting the requirements of the Directive.

The manufacturer, under the R&TTE Directive, is required to provide the user of his product with information about its use, and also a declaration of conformity with the essential requirements of the Directive. The product packaging and instructions must bear the CE mark, the number of the notified body that was consulted to get guidance on test requirements and meeting the essential requirements, and, if there are operating restrictions because of use of nonharmonized frequency bands, an "alert sign" consisting of an exclamation point contained within a circle. It shall also be identified by a serial number, or type or batch number, and the name of the manufacturer or the person responsible for placing the product on the market. When the alert sign is used, the manufacturer is responsible for notifying the states in the EU where operation of his product is restricted.

The manufacturer has a choice of several routes for demonstrating compliance with the requirements of the Directive. In the case of radio equipment, the procedures are specified in Annexes III, IV, and V, outlined below. When the manufacturer applies the harmonized standards that are appropriate to his product, he may choose any of the procedures in those Annexes. However, if he does not apply harmonized standards or applies them only in part, he can choose between following the procedures of Annexes IV or V.

Annex III

The manufacturer prepares technical documentation covering design, manufacture and operation of the product and including the following details:

- A general description of the product, conceptual design and manu-facturing drawings and schemes of components, sub-assemblies, circuits, etc.,

- Descriptions and explanations necessary for the understanding of said drawings and schemes and the operation of the product,

- A list of the standards referred to in Article 5 (harmonized stan-dards), applied in full or in part, and descriptions and explanations of the solutions adopted to meet the essential requirements of the Directive where such standards referred to in Article 5 have not been applied or do not exist,

- Results of design calculations made, examinations carried out, etc.,

- Test reports.

The manufacturer performs the test suites specified by the relevant harmonized standards. If the test suites are not given in the standards, then he chooses a notified body to identify the test suites for him. Notified bodies are designated by national authorities responsible for telecommunication matters to perform certain tasks under the procedures of the Directive when the manufacturer does not apply harmonized standards in full. The manufacturer, or the person responsible for putting the goods on the market, declares that the tests were carried out and that the product complies with the essential requirements. Marking on the product includes the CE Mark and the number of the notified body, if consulted.

Annex IV

This procedure is similar to that of Annex III, but in addition, the manufacturer prepares a technical construction file (TCF) consisting of the documentation detailed under Annex III above along with a declaration of conformity to specific radio test suites. The TCF is presented to a notified body who gives an opinion on whether the requirements of the Directive are met. When placing the product on the market, the manufacturer includes the number of the notified body after the CE mark on the device and packaging, as prescribed.

Annex V

The manufacturer operates an approved quality system for design, manufacture and final product inspection and testing which is subject to surveillance by a notified body. The notified body must assess whether the quality control system ensures conformity of the products with the requirements of the Directive in light of relevant documentation and test results supplied by the manufacturer. When placing his product on the market, the manufacturer includes the number of the notified body after the CE mark on the device and packaging, as prescribed.

It should be stressed that the R&TTE Directive regulates the putting of telecommunication, including wireless, devices on the market and allows their distribution generally throughout all European countries belonging to CEPT. However, operating characteristics, including frequency bands, maximum power output and duty cycle, for example, as well as licensing requirements, are still determined by the authorities of the individual countries. Thus, the manufacturer affixes the "alert sign" on his product when those characteristics are not fully harmonized and is responsible for indicating to the user where the device may be used. The ERO web site, listed in the "References and Bibliography" section should be referred to for up-to-date information on the degree of harmonization of transmission characteristics in the European countries.

As an example of the application of the R&TTE Directive, we'll describe the relevant harmonized documents which are the basis of testing compliance in the case of unlicensed short-range devices. The harmonized document for the radio transmission characteristics is EN 300 220-3. This is part 3 of the specification, EN 300 220-1, described previously, designated as "Harmonized EN covering essential requirements under article 3.2 of the R&TTE Directive." It includes technical requirements and lists the essential test suites to be used for compliance with the directive while referring to the detailed document EN 300 220-1.

Documents EN 301 489-1 and EN 301 489-3 are used to confirm compliance with the electromagnetic compatibility requirements of the Directive. These requirements involve both the potential for interfering with other equipment through radiation, and the susceptibility to being interfered with by other equipment or phenomena, or immunity to interference. The radiation part of the requirement generally covers nonintentional radiation, from digital control circuits for example, whereas intentional radiation and its spurious by-products are covered by the transmission characteristic specifications mentioned in the paragraph above. Phenomena which effect the interference immunity of the equipment being tested are

- Electromagnetic fields in the range 80 to 1000 MHz
- Electrostatic discharge
- Fast transients on signal, control and power ports
- RF common mode interference—.15 to 80 MHz
- Transients and surges on DC power input ports
- AC power line variations and anomalies

EN 301 489-3 sets out exclusion bands for receivers, based on the device's intended function, to prevent susceptibility to interference on an intended received frequency, or image frequency, for example, from disqualifying the product. For transmitters, radiation on the operating frequency and sidebands is excluded from the interfering possibilities of the device.

The essential requirements regarding health and safety, referenced in article 3 paragraph 1(a) of the Directive, are satisfied by meeting the requirements in the CENELEC (European committee for electrotechnical standardization) document EN 60950, "Safety of information technology equipment." The nature of testing involved depends on whether the product has an internal or external power supply unit.

A designer of short-range radio equipment that is to be marketed in the European Union is well advised to take into account the EMC and safety requirements of the R&TTE Directive early in the development stages of the product.

9.5 Standards in the United Kingdom

Short-range devices for alarm systems and telecontrol are used on frequencies other than those listed under ERC 70-3 (Table 1). UK regulations for short-range devices which were numbered MPT 13xx have been superseded by EN 300 220. Some UK frequency allocations and their characteristics are:

- 49.82 to 49.98 MHz for general purpose use, up to 10 mW ERP, bandwidth 10 kHz or wide-band.

- 173.225 MHz for fixed or short-range alarms with 12.5 or 25 kHz bandwidth and up to 10 mW ERP.

- 418 MHz frequency for alarm systems. Maximum power allowed is 250 µW. This frequency has been very popular in the UK in the past because the common European frequency, 433 MHz, was allowed only for automotive keyless entry. However, in 1998 the UK aligned itself with the ERC 70-3E recommendations regarding 433 MHz, and 418 MHz may no longer be used for new equipment. One reason is that a relatively high-power mobile telephone trunk service called TETRA is coming into use in the UK between 410 and 430 MHz and will potentially interfere with 418 MHz systems.

- 458.825 MHz for alarms with up to 100 mW and 12.5 kHz bandwidth.

Details and updates concerning short-range devices (low power devices) are available on the Radiocommunications Agency web site.

9.6 Japanese Low Power Standards

In general, the radio regulations in Japan that pertain to short-range wireless are quite different from those in North America and Europe. Devices certified for marketing and operation in the West will not meet the requirements for the same type of function in Japan. The popular industrial standards, Bluetooth and Wi-Fi, do contain provisions for meeting the 2.4 MHz band regulations in Japan, but hopping patterns and frequency channels are unique so the products must be specially designed for use in that country.

We will review three aspects of Japanese regulations that apply to unlicensed short-range wireless communication:

- Very low power transmitters

- Telemetry and security transmissions in 400 and 1200 MHz bands

- WLAN and WPAN communications on the 2.4 GHz band

Very low power transmitters

This class applies to what are called "extremely weak power equipment." Examples are very low output cordless telephones, wireless microphones, emergency alarm transmitters, and toys. Requirements are the least restricted of the unlicensed regulations and in this sense they may be compared to Section 15.209 of the FCC regulations. Allowed field strength is higher, up to 322 MHz, but there are no provisions for even higher power, comparable to FCC Section 15.231. Allowed field strength as a function of frequency, up to 1000 GHz, is shown in Figure 9-4. Measurement distance is three meters. The limit is 500 µV/m up to 322 MHz and 35 µV/m from there up to 10 GHz.

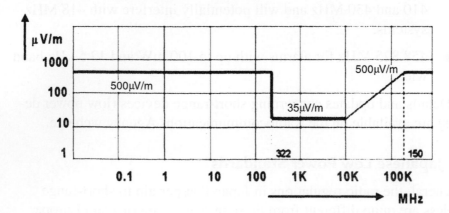

Figure 9-4. Japanese Very Low Power Field Strength Limits

Telemetry and Security

Radio telemetry and data transmission functions (primarily for processing by machines) are provided for over narrow band channels in 400 and 1200 MHz frequency bands, with power limits of 1 and 10 milliwatts. A detailed specification in English, ARIB STD-67T, is available from the Association of Radio Industries and Businesses (ARIB). ARIB standards set forth but also complement the mandatory requirements of the Japanese Ministry of Public Management, Home Affairs, Post and Telecommunications regulations. They include "private technical standards" that aim to provide compatibility of radio facilities and transmission quality along with greater convenience to radio equipment manufacturers and users.

The principle characteristics and limits from the telemetry and data transmission technical requirements are given in the following table. More details, including recommendations for duplex frequency pairing and receiver requirements, are given in ARIB STD-67T.

Band No.	Frequency MHz	Channel Separation kHz	Occupied BW kHz	Average Power to Antenna mW
1	426.0250-426.1375	12.5	8.5	1
2	429.1750-429.2375	12.5	8.5	10
3	429.2500-429.7375	12.5	8.5	10
4	426.0375-426.1125	25	16	1
5	429.8125-429.9250	12.5	8.5	10
6	449.7125-449.8875	12.5	8.5	10
7	469.4375-469.4875	12.5	8.5	10
8	1216.0125-1216.9875	25	16	10
9	1252.0125-1252.9875	25	16	10
10	1216.0000-1217.0000	50	32	10
11	1252.0000-1253.0000	50	32	10

Note that some bands overlap.

Other characteristics:

- Modulation: Frequency or phase, or combination of amplitude and frequency or phase
- Frequency tolerance: ±4 ppm (±3 ppm for bands 8,9)

- Adjacent channel power: −40 dB down from carrier

- Spurious emissions, including harmonics: 2.5 microwatts

- Receiver radiation (to receiver antenna): 4 nanowatts

- Carrier sensing capability and transmission time restrictions specified for particular channels

The band of frequencies from 426.25 to 426.8375 MHz is allocated for low-power security radio systems. Channel separation is 12.5 or 25 kHz, depending on occupied bandwidth that may be from 4 to 16 kHz. Most characteristics are the same as those above, except that a frequency tolerance of ±10 ppm is allowed for occupied bandwidth of 4 kHz and 12.5 kHz channel separation, and occupied bandwidth of 8.5 to 12 kHz for channel separation of 25 kHz. Stations must automatically send and receive an identification code. Transmissions are required to cease after three seconds and wait at least two seconds before transmitting again.

WLAN and WPAN on 2.4 GHz

Both Bluetooth and Wi-Fi products can be used in Japan, although with limitations that are referred to in the respective IEEE industrial standards, 802.15.1 and 802.11. The available spectrum for the unlicensed wireless networks in Japan is 2.473 to 2.495 GHz, compared to 2.4 to 2.483 MHz in North America and Europe (there are exceptions in some European countries). Bluetooth operation in Japan is limited to 23 hopping channels, compared to 79 in most other countries. For Wi-Fi, there is only one channel, centered on 2.484 GHz.

9.7 Non-Governmental Standards

The sections above discuss the various mandatory requirements for short-range radio devices in the US and Canada, Europe and Japan. However, there are various consumer interest organizations that have issued specifications for such equipment that, although not legally imposing, can affect the ability of a manufacturer to sell his product.

U.S.A.—Underwriters Laboratories

Underwriters Laboratories standard UL 1023 "Household Burglar-Alarm System Units" includes requirements for short-range radio frequency devices that "…are applicable to control units and systems that utilize initiating, annunciating, and remote control devices that are not interconnected by a solid medium…"

We give a cursory review of some of the parameters that are tested, without going into details of the test procedures or limits.

a) *Interference immunity*. Signals from an interfering carrier modulated in turn by white noise, audio frequency between 20 and 40,000 Hz, and a square wave between 20 and 40,000 Hz shall not cause false alarming. The interfering signal frequency is varied ±5% about the desired carrier frequency and its image and IF frequencies if applicable.

b) *Frequency selectivity*. A product's receiver shall not respond to a strong (defined) system transmitter operating on a frequency more than two channel widths above and below the desired center frequency.

c) *Clash rate*. Specifies the probability of a loss of alarm information due to two or more transmitters being concurrently activated.

d) *Error rate*. Specifies the measure of the ability of a receiver to discriminate between correct and incorrect transmissions so that erroneous signals are not accepted as valid status indications from the various transmitters.

e) *Throughput rate*. Specifies a measure of the ability of a receiver to accurately interpret an alarm or status signal in order to achieve a high degree of assurance that alarm or emergency signals are not lost.

UL 1023 also prescribes transmitter stability tests over extreme temperatures and an accelerated aging test.

UK Consumer Standards

British Standard 6799 defines five classes of wireless security systems. Distinctions between the classes involve low battery reporting, supervision, jamming detection and the number of distinguishing codes. Following are the features of each class: Each class (except the first) also has the features of the previous classes.

- Class I: Gives local or remote indication of a low battery condition at least seven days in advance of battery failure.

- Class II: Provides indication at the receiver of the transmitter that operated.

- Class III: Generates fault indication at the receiver if a receiver is blocked by interference or jamming for more than 30 seconds.

- Class IV: Provides return to normal signal from detector when the cause of an alarm ceases. Also, sends supervision signals at periodic intervals not exceeding 8.4 hours and creates alarm if three successive supervision signals are not received. Transmits low battery signal to receiver and generates fault condition if low battery signals continue for more than 8.4 hours.

- Class V: Sends supervision signals at intervals up to 1.2 hours and generates alarm if no supervision signal is receiver from any particular transmitter within 3.6 hours.

The National Approval Council for Security Systems (NACOSS) has a Code of Practice for wireless systems which it approves. It is based on BS 6977 with some modification, and a sixth class with more severe requirements. NACOSS recognizes, inspects and regulates firms who install, maintain and monitor electronic security systems.

Appendix 9-A

Terms and Definitions (FCC Part 2)

Assigned Frequency Band. The frequency band within which the emission of a station is authorized; the width of the band equals the necessary bandwidth plus twice the absolute value of the frequency tolerance. Where space stations are concerned, the assigned frequency band includes twice the maximum Doppler shift that may occur in relation to any point of the Earth's surface. Assignment (of a radio frequency or radio frequency channel). Authorization given by an administration for a radio station to use a radio frequency or radio frequency channel under specified conditions.

Carrier Power (of a radio transmitter). The average power supplied to the antenna transmission line by a transmitter during one radio frequency cycle taken under the condition of no modulation.

Characteristic Frequency. A frequency which can be easily identified and measured in a given emission. NOTE: A carrier frequency may, for example, be designated as the characteristic frequency.

Class of Emission. The set of characteristics of an emission, designated by standard symbols, e.g., type of modulation, modulating signal, type of information to be transmitted, and also if appropriate, any additional signal characteristics.

Direct Sequence Systems. A spread-spectrum system in which the carrier has been modulated by a high-speed spreading code and an information data stream. The high-speed code sequence dominates the "modulating function" and is the direct cause of the wide spreading of the transmitted signal.

Duplex Operation. Operating method in which transmission is possible simultaneously in both directions of a telecommunication channel.

Effective Radiated Power (e.r.p) (in a given direction). The product of the power supplied to the antenna and its gain relative to a halfwave dipole in a given direction.

Equivalent Isotropically Radiated Power (e.i.r.p.). The product of the power supplied to the antenna and the antenna gain in a given direction relative to an isotropic antenna.

Equivalent Monopole Radiated Power (e.m.r.p.) (in a given direction). The product of the power supplied to the antenna and its gain relative to a short vertical antenna in a given direction.

Frequency-hopping Systems. A spread-spectrum system in which the carrier is modulated with the coded information in a conventional manner causing a conventional spreading of the RF energy about the frequency carrier. The frequency of the carrier is not fixed but changes at fixed intervals under the direction of a coded sequence. The wide RF bandwidth needed by such a system is not required by spreading of the RF energy about the carrier but rather to accommodate the range of frequencies to which the carrier frequency can hop. The test of a frequency-hopping system is that the near term distribution of hops appears random, the long term distribution appears evenly distributed over the hop set, and sequential hops are randomly distributed in both direction and magnitude of change in the hop set.

Frequency Tolerance. The maximum permissible departure by the center frequency of the frequency band occupied by an emission from the assigned frequency or, by the characteristic frequency of an emission from the reference frequency.
NOTE: The frequency tolerance is expressed in parts in 10^6 or in hertz.

Gain of an Antenna. The ratio, usually expressed in decibels, of the power required at the input of a loss free reference antenna to the power supplied to the input of the given antenna to produce, in a given direction, the same field strength or the same power flux-density at the same distance. When not specified otherwise, the gain refers to the direction of maximum radiation. The gain may be considered for a specified polarization.

Harmful Interference. Interference which endangers the functioning of a radio navigation service or of other safety services or seriously degrades, obstructs, or repeatedly interrupts a radiocommunication service operating in accordance with these [international] Radio Regulations.

Hybrid Spread-Spectrum Systems. Hybrid spread-spectrum systems are those which use combinations of two or more types of direct sequence, frequency hopping, time hopping and pulsed FM modulation in order to achieve their wide occupied bandwidths.

Industrial, Scientific and Medical (ISM) (of radio frequency energy) Applications. Operation of equipment or appliances designed to generate and use locally radio frequency energy for industrial, scientific, medical, domestic or similar purposes, excluding applications in the field of telecommunications.

Interference. The effect of unwanted energy due to one or a combination of emissions, radiations, or inductions upon reception in a radio communication system, manifested by any performance degradation, misinterpretation, or loss of information which could be extracted in the absence of such unwanted energy.

Necessary Bandwidth. For a given class of emission, the width of the frequency band which is just sufficient to ensure the transmission of information at the rate and with the quality required under specified conditions.

Occupied Bandwidth. The width of a frequency band such that, below the lower and above the upper frequency limits, the mean powers emitted are each equal to a specified percentage Beta/2 of the total mean power of a given emission.
NOTE: Unless otherwise specified by the CCIR for the appropriate class of emission, the value of Beta/2 should be taken as 0.5%.

Out-of-band Emission. Emission on a frequency or frequencies immediately outside the necessary bandwidth which results from the modulation process, but excluding spurious emissions.

Peak Envelope Power (of a radio transmitter). The average power supplied to the antenna transmission line by a transmitter during one radio frequency cycle at the crest of the modulation envelope taken under normal operating conditions.

Pulsed FM Systems. A pulsed FM system is a spread-spectrum system in which a RF carrier is modulated with a fixed period and fixed duty cycle sequence. At the beginning of each transmitted pulse, the carrier frequency is frequency modulated causing an additional spreading of the carrier. The pattern of the frequency modulation will depend upon the spreading function which is chosen. In some systems the spreading function is a linear FM chirp sweep, sweeping either up or down in frequency.

Radiation. The outward flow of energy from any source in the form of radio waves.

Radio. A general term applied to the use of radio waves.

Radiocommunication. Telecommunication by means of radio waves.

Semi-Duplex Operation. A method which is simplex operation at one end of the circuit and duplex operation at the other.

Simplex Operation. Operating method in which transmission is made possible alternatively in each direction of a telecommunication channel, for example, by means of manual control.

Spread-Spectrum Systems. A spread spectrum system is an information bearing communications system in which: (1) Information is conveyed by modulation of a carrier by some conventional means, (2) the bandwidth is deliberately widened by means of a spreading function over that which would be needed to transmit the information alone. (In some spread spectrum systems, a portion of the information being conveyed by the system may be contained in the spreading function.)

Spurious Emission. Emission on a frequency or frequencies which are outside the necessary bandwidth and the level of which may be reduced without affecting the corresponding transmission of information. Spurious emissions include harmonic emissions, parasitic emissions, intermodulation products and frequency conversion products, but exclude out-of-band emissions.

Telecommand. The use of telecommunication for the transmission of signals to initiate, modify or terminate functions of equipment at a distance.

Telecommunication. Any transmission, emission or reception of signs, signals, writing, images and sounds or intelligence of any nature by wire, radio, optical or other electromagnetic systems.

Telemetry. The use of telecommunication for automatic indicating or recording measurements at a distance from the measuring instrument.

Appendix 9-B

Nomenclature for Defining Emission, Modulation and Transmission (FCC Part 2)

The system for defining radio signals is based on the Radio Regulations published by the International Telecommunications Union. A basic description is given below. For more information see FCC regulations, paragraphs 2.201 and 2.202, or the Radiocommunication Authority of the UK document RA97.

Emissions are defined by the necessary bandwidth (see definition in Appendix 9-A) and a three-character Class of Emission. As an example, "12K5F3E" represents an analog FM telephony signal with a necessary bandwidth of 12.5 kHz.

Necessary Bandwidth

The necessary bandwidth is expressed by three numerals and one letter. The letter occupies the position of the decimal point and represents the unit of bandwidth as follows:

From 0.001 to 999 Hz = H

From 1.00 to 999 kHz = K

From 1.00 to 999 MHz = M

From 1.00 to 999 GHz = G

Note: The first character shall be neither zero nor K, M or G.

Examples:

1 Hz = 1H00

400 Hz = 400H

6 KHz = 6K00

10 MHz =10M0

Class of Emission

The basic characteristics are

first character: type of modulation of the main carrier
second character: nature of signal modulating the main carrier
third character: type of information to be transmitted

First Character

1) N = Unmodulated carrier

2) Amplitude Modulation

A = double sideband

H = single sideband, full carrier

R = single sideband, reduced or variable level carrier

J = single sideband, suppressed carrier

B = independent side-bands

C = vestigial sideband

3) Angle Modulation

F = frequency modulation

G = phase modulation

D = amplitude and angle modulation either simultaneously or in a pre-established sequence

Note: Emissions where the main carrier is directly modulated by a signal that has been coded into quantized form, such as pulse code modulation, are designated under the sections above. Other types of pulse modulation are represented by letters not given here but are listed in the sources referred to above.

Second Character

0 = no modulating signal

1 = a single channel containing quantized or digital information without the use of a modulating subcarrier. This excludes time-division multiplex.

2 = a single channel containing quantized or digital information with the use of a modulating subcarrier. This excludes time division multiplex.

3 = a single channel containing analog information

7 = two or more channels containing quantized or digital information

8 = two or more channels containing analog information

9 = composite system with one or more channels containing analog quantized or digital information, together with one or more channels containing analog information

X = cases not otherwise covered

Third Character

N = no information transmitted

A = telegraphy for aural reception

B = telegraphy for automatic reception

C = facsimile

D = data transmission, telemetry, telecommand

E = telephony (including sound broadcasting)

F = television (video)

W = combination of the above

X = cases not otherwise covered

Appendix 9-C

Restricted Frequencies and Field Strength Limits from Section 15.205 of FCC Rules and Regulations

Section 15.205 <u>Restricted bands of operation</u>.

(a) Except as shown in paragraph (d) of this section, only spurious emissions are permitted in any of the frequency bands listed below:

MHz	*MHz*	*MHz*	*GHz*
0.090 – 0.110	16.42 – 16.423	399.9 – 410	4.5 – 5.15
(^{1}) .495 – 0.505	16.69475 –	608 – 614	5.35 – 5.46
2.1735 – 2.1905	16.69525	960 – 1240	7.25 – 7.75
4.125 – 4.128	16.80425 –	1300 – 1427	8.025 – 8.5
4.17725 – 4.17775	16.80475	1435 – 1626.5	9.0 – 9.2
4.20725 – 4.20775	25.5 – 25.67	1645.5 – 1646.5	9.3 – 9.5
6.215 – 6.218	37.5 – 38.25	1660 – 1710	10.6 – 12.7
6.26775 – 6.26825	73 – 74.6	1718.8 – 1722.2	13.25 – 13.4
6.31175 – 6.31225	74.8 – 75.2	2200 – 2300	14.47 – 14.5
8.291 – 8.294	108 – 121.94	2310 – 2390	15.35 – 16.2
8.362 – 8.366	123 – 138	2483.5 – 2500	17.7 – 21.4
8.37625 – 8.38675	149.9 – 150.05	2655 – 2900	22.01 – 23.12
8.41425 – 8.41475	156.52475 –	3260 – 3267	23.6 – 24.0
12.29 – 12.293	156.52525	3332 – 3339	31.2 – 31.8
12.51975 –	156.7 – 156.9	3345.8 – 3358	36.43 – 36.5
12.52025	162.0125 – 167.17	3600 – 4400	(^{2})
12.57675 –	167.72 – 173.2		
12.57725	240 – 285		
13.36-13.41	322 – 335.4		

(1) Until February 1, 1999, this restricted band shall be 0.490–0.510 MHz.
(2) Above 38.6.

(b) Except as provided in paragraphs (d) and (e), the field strength of emissions appearing within these frequency bands shall not exceed the limits shown in Section 15.209. At frequencies equal to or less than 1000 MHz, compliance with the limits in Section 15.209 shall be demonstrated using measurement instrumentation employing a CISPR quasi-peak detector. Above 1000 MHz, compliance with the emission limits in Section 15.209 shall be demonstrated based on the average value of the measured emissions. The provisions in Section 15.35 apply to these measurements.

(c) Except as provided in paragraphs (d) and (e), regardless of the field strength limits specified elsewhere in this Subpart, the provisions of this Section apply to emissions from any intentional radiator.

CHAPTER 10

Introduction to Information Theory

Up to now, all the chapters in this book have related directly to radio communication. This chapter is different. Information theory involves communication in general—on wires, fibers, or over the air—and it's applied to widely varied applications such as information storage on optical disks, radar target identification, and the search for extraterrestrial intelligence. In general, the goal of a communication system is to pass "information" from one place to another through a medium contaminated by noise, at a particular rate, and at a minimum specified level of fidelity to the source. Information theory gives us the means for quantitatively defining our objectives and for achieving them in the most efficient manner. The use of radio as a form of communication presents obstacles and challenges that are more varied and complex than a wired medium. A knowledge of information theory lets us take full advantage of the characteristics of the wireless interface.

In order to understand what information theory is about, you need at least a basic knowledge of probability. We start this chapter with a brief exposition of this subject. We've already encountered uses of probability theory in this book—in comparing different transmission protocols and in determining path loss with random reflections.

The use of coding algorithms is very common today for highly reliable digital transmission even with low signal-to-noise ratios. Error correction is one of the most useful applications of information theory.

Finally, information theory teaches us the ultimate limitations in communication—the highest rate that can be transmitted with a given bandwidth and a given signal-to-noise ratio.

10.1 Probability

Basics

A common use of probability theory in communication is in assessing the reliability of a received message transmitted over a noisy channel. Let's say we send a digital message frame containing 32 data bits. What is the probability that the message will be received in error—that is, that one or more bits will be corrupted? If we are using or are considering using an error-correcting code that can correct one bit, then we will want to know the probability that two or more bits will be in error. Another interesting question: what is the probability of error of a frame of 64 bits, compared to that of 32 bits, when the probability of error of a single bit is the same in both cases?

In all cases we must define what is called an *experiment* in probability theory. This involves defining *outcomes* and *events* and assigning probability measures to them that follow certain rules.

We state here the three axioms of probability, which are the conditions for defining probabilities in an experiment with a finite number of outcomes. We also must describe the concept of *field* in probability theory. Armed with the three axioms and the conditions of a field, we can assign probabilities to the events in our experiments. But, first, let's look at some definitions.

An *outcome* is a basic result of an experiment. For example, throwing a die has six outcomes, each of which is a different number of dots on the upper face of the die when it comes to rest. An *event* is a set of one or more outcomes that has been defined, again according to rules, as a useful observation for a particular experiment. For an example, one event may be getting an odd number from a throw of a die and another event may be getting an even number. The outcomes are assigned probabilities and the events receive probabilities from the outcomes that they are made up of, in accordance with the three axioms. The term *space* refers to the set of all of the outcomes of the experiment. We'll define other terms and concepts as we go along, after we list the three axioms.

Axioms of probability

    I. $P(A) \geq 0$ The probability of an event A is zero or positive.

    II. $P(S) = 1$ The probability of space is unity.

    III. If $A \cdot B = 0$ then $P(A + B) = P(A) + P(B)$ If two events A and B are mutually exclusive, then the probability of their sum equals the sum of their individual probabilities.

The product of two events, shown in Axiom III as $A \cdot B$ and often called *intersection,* is an event which contains the outcomes that are common to the two events. The sum of two events $A + B$, also called *union*, is an event which contains all of the outcomes in both component events.

Mutually exclusive, referred to in Axiom III, means that two events have no outcomes in common. This means that if in an experiment one of the events occurs, the other one doesn't. Returning to the experiment of throwing a die, the odd event and the even event are mutually exclusive.

Now we give the definition of a field, which tells us what events we must include in an experiment. We will use the term *complement* of an event, which is all of the outcomes in the space not included in the event. The complement of A is A'. $A \cdot A' = 0$.

Definition of a field F

    1. If $A \in F$ then $A' \in F$

        If event A is contained in the field F, then its complement is also contained in F.

    2. If $A \in F$ and $B \in F$ then $A + B \in F$
        If the events A and B are each contained in F, then the event that is the sum of A and B is also contained in F.

Now we need one more definition before we get back down to earth and deal with the questions raised and the uses mentioned at the beginning of this section.

Definition of independent events

Two events are called *independent* if the probability of their product (intersection) equals the product of their individual probabilities:

$$P(A \cdot B) = P(A) \times P(B)$$

This definition can be extended to three or more events. For three independent events A, B, C:

$$P(A \cdot B \cdot C) = P(A) \times P(B) \times P(C) \text{ and}$$

$$P(A \cdot B) = P(A) \times P(B); P(A \cdot C) = P(A) \times P(C); P(B \cdot C) = P(B) \times P(C)$$

Similarly, for more than three events the probability of the product of all events equals the product of the probabilities of each of the events, and the probability of the product of any lesser number of events equals the product of the probabilities of those events. If there are n independent events, then the total number of equations like those shown above for three events that are needed to establish their independence is $2^n - (n+1)$.

Examples

We now can look at some examples of how to use probability theory.

Problem: What is the probability of correctly receiving a sequence of 12 bits if the probability of error of a bit is one out of one hundred, or $p_e = .01$? All bits are independent.

Solution: We look at the problem as an experiment in which we must define space, events according to the conditions of a field, and the probabilities of the events. We'll call the probability of a correct sequence P_c and the probability of an error in the sequence P_e.

The space in our experiment contains all conceivable outcomes. Since we have a sequence of 12 bits, we can receive $2^{12} = 4096$ different sequences of bits, or words. The events in our experiment, conforming to the conditions for a field, are

(1) The reception of the correct word—that is, no bits are in error.

(2) The reception of an erroneous word—a sequence that has 1 or more bits in error.

(3) The reception of any word.

(4) The reception of no word.

The inclusion of event (3) is necessary because of condition 2 in the definition of a field, which says that the event that is the sum of events must also be in the field. The sum of events (1) and (2) is all of the outcomes, which is the space, and this is event (3). Event (4) is needed because of condition 1 of a field—the complement of any event must be included—and event (4) is the complement of event (3). The complements of events (1) and (2) are each other, so the requirements of a field are complied with.

Now we assign probabilities to the events. In the statement of the problem we designated the probability of a bit error as p_e. In the field of an individual bit, we have two events: bit error or no bit error. The sum of these two events is the bit space, whose probability is unity. It follows that the probability of no bit error + probability of bit error (p_e) = probability of space (1). Thus, the probability of no bit error = $1 - p_e$.

Now, the bits in the received sequence are independent, so the probability of a particular sequence equals the product of the probabilities of each of its bits. In the case of the errorless sequence, the probability that each bit has no bit error is $1 - p_e$, so this sequence's probability is $(1 - p_e)^{12}$. Using the given bit error probability of .01, we find that the probability of correctly receiving the sequence is $.99^{12}$ or approximately 88.6 percent.

How can we interpret this answer? If the sequence is sent only once for all time, it will either be received correctly or it won't. In this case, the establishment of a probability won't have much significance, except for the purpose of placing bets. However, if sequences are sent repeatedly, we will find that as the number of sequences increases, the percentage of those correctly received approaches 88.6.

Just as we found the probability of no bit error = $1 - p_e$, we find the probability of a sequence error is $1 - .886 = 0.114$. This is from axioms II and III and the fact that the sum of the two mutually exclusive events—incorrect and correct sequences—is space, whose probability is 1.

Figure 10-1 gives a visual representation of this problem, showing space, the events, and the outcomes. Each outcome, representing one of the 4096 sequences, is assigned a probability P_i :

$$P_i = p_1 p_2 p_3 p_4 p_5 p_6 p_7 p_8 p_9 p_{10} p_{11} p_{12}$$

where p_1, p_2, and so on equals either p_e or $1 - p_e$, depending on whether that specific bit in the sequence is in error or not. Sequences having the same number of error bits have identical probabilities.

For example, an outcome that is a sequence having one bit in error has a probability of $p_e(1-p_e)^{11}$. There are 12 of these mutually exclusive sequences, so the probability of receiving a sequence having one and only one error bit is

$$P_1 = 12\, p_e(1-p_e)^{11} = .107$$

Conditional Probability

An important concept in probability theory is *conditional probability*. It is defined as follows:

Given an event B with nonzero probability $P(B) > 0$, then the conditional probability of event A assuming event B is known as:

$$P(A|B) = P(A \cdot B)/P(B) \tag{10-1}$$

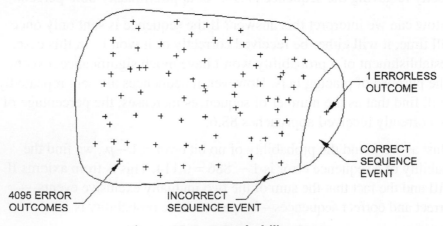

Figure 10-1: A Probability Space

A consequence of this definition is that the probability of an event is affected by the assumption of occurrence of another event. We see from the expression above that if A and B are mutually exclusive, the conditional probability is zero because $P(A \cdot B) = 0$. If A and B are independent, then the occurrence of B has no effect on the probability of $A : P(A|B) = P(A)$. Conditional probabilities abide by the three axioms and the definition of a field.

Up to now we have discussed only sets of finite outcomes, but the theory holds when events are defined in terms of a continuous quantity as well. A space can be the set of all real numbers, for example, and subsets, or events, to which we can assign probabilities are intervals in this space. For example, we can talk about the probability of a train arriving between 1 and 2 p.m., or of the probability of a received signal giving a detector output above 1 volt. The axioms and definition of a field still hold but we have to allow for the existence of infinite sums and products of events.

Density and Cumulative Distribution Functions

In the problem above we found the probability, in a sequence of 12 bits, of getting no errors, of getting an error (one or more errors), and of getting an error in one bit only. We may be interested in knowing the probability of receiving exactly two bits in error, or any other number of error bits. We can find it using a formula called the binomial distribution:

$$P_m(n) = \binom{m}{n} p^n q^{m-n} \tag{10-2}$$

where $P_m(n)$ is the probability of receiving exactly n bits in error, m is the total number of bits in the sequence, p is the probability of one individual bit being in error and $q = 1 - p$, the probability that an individual bit is correct. $\binom{m}{n}$ represents the number of different combinations of n objects taken m at a time:

$$\binom{m}{n} = \frac{m!}{n!(m-n)!}$$

The notation ! is factorial—for example, $m! = m(m-1)(m-2) \ldots (1)$.

Each time we send an individual sequence, the received sequence will have a particular number of bits in error—from 0 to 12 in our example. When a large number of sequences are transmitted, the frequency of having exactly n errors will approach the probability given in Eq. (10-2). In probability theory, the quantity that expresses the observed result of a random process is called a *random variable*. In our example, we'll call this random variable N. Thus, we could rewrite Eq. (10-2) as:

$$P_m\left(N=n\right)=\binom{m}{n}p^n q^{m-n} \tag{10-2a}$$

(We represent uppercase letters as random variables and lowercase letters as real numbers. We may write $P(x)$ which means the probability that random variable X equals real number x.)

The random value can be any number that expresses an outcome of an experiment, in this case 0 through 12. The random values are mutually exclusive events.

The probability function given in Eq. (10-2a), which gives the probability that the random variable equals a discrete quantity, is sometimes called the *frequency function*. Another important function is the *cumulative probability distribution function*, or *distribution function*, defined as

$$F(x) = \text{Prob}(X \le x) \tag{10-3}$$

defined for any number x from $-\infty$ to $+\infty$ and X is the random value.

In our example of a sequence of bits, the distribution function gives the probability that the sequence will have n or fewer bits in error, and its formula is

$$F\left(n\right)=\sum_{n=0}^{m}\binom{m}{n}p^n q^{m-n} \tag{10-4}$$

where Σ is the symbol for summation.

The example which we used up to now involves a discrete random variable, but probability functions also relate to continuous random variables, such as time or voltage. The best known of these is probably the Gaussian probability function, which describes, for example, thermal noise in a radio receiver. The analogous type of function to the frequency probability function defined for the discrete variable is called a density function. The Gaussian probability density function is:

$$P(x) = \frac{1}{\sqrt{2 \cdot \pi \cdot \sigma^2}} \cdot e^{\frac{-(x-m)^2}{2\sigma^2}}$$

(10-5)

where σ^2 is the variance and m is the average (defined below).

Plots of the frequency function $P_m(n)$ (p and $q = 1/2$) and the density function $P(x)$, with the same average and variance, are shown in Figure 10-2. While these functions are analogous, as stated above, there are also fundamental differences between them. For one, $P(x)$ is defined on the whole real abscissa, from $-\infty$ (infinity) to $+\infty$ whereas $P_m(n)$ can have only the discrete values 0 to 12. Second, points on $P(x)$ are not probabilities! A continuous random variable can have a probability greater than zero only over a finite interval. Thus, we cannot talk about the probability of an instantaneous noise voltage of 2 volts, but we can find the probability of it being between, say, 1.8 and 2.2 volts. Probabilities on the curve $P(x)$ are the area under the curve over the interval we are interested in. We find these areas by integrating the density curve between the endpoints of the interval, which may be plus or minus infinity.

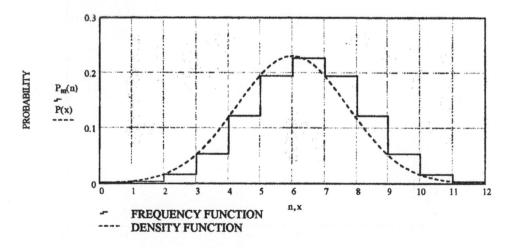

Figure 10-2: Frequency and Density Functions

The more useful probability function for finding probabilities directly for continuous random variables is the distribution function. Figure 10-3 shows the Gaussian distribution function, which is the integral of the density function between $-\infty$ and x. All distribution functions have the characteristics of a positive slope and values of 0 and 1 at the extremities. To find the probability of a random variable over an interval, we subtract the value of the distribution function evaluated at the lower boundary from its value at the upper boundary. For example, the probability of the interval of 4 to 6 in Figure 10-3 is

$$F(6) - F(4) = .5 - .124 = .376$$

Average, Variance, and Standard Deviation

While the distribution function is easier to work with when we want to find probabilities directly, the density function or the frequency function is more convenient to use to compute the statistical properties of a random variable. The two most important of these properties are the *average* and the *variance*.

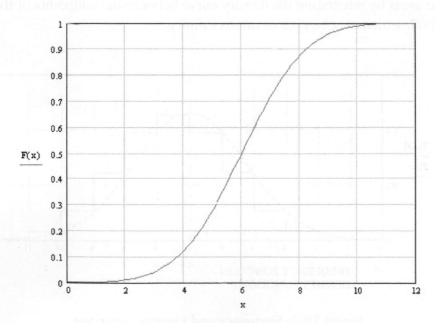

Figure 10-3: Gaussian Distribution Function

The statistical average for a discrete variable is defined as

$$\overline{X} = \sum_i x_i \cdot P(x_i)$$

(10-6)

Writing this using the frequency function for the *m*-bit sequence (Eq. (10.2)) we have:

$$\overline{X} = \sum_n n \cdot P_m(n)$$

where *n* ranges from 0 to *m*.

Calculating the average using p = .1, for example, we get $\overline{X}$ = 1.2 bits. In other words, the average number of bits in error in a sequence with a bit error probability of .1 is just over 1 bit.

The definition of the average of a continuous random variable is

$$\overline{X} = \int_{-\infty}^{+\infty} x \cdot P(x)dx$$

(10-7)

If we apply this to the expression for the Gaussian density function in Eq. (10-5) we get $\overline{X} = m$ as expected.

We can similarly find the average of a function of a random variable, in both the discrete and the continuous cases:

When this function is expressed as $f(x) = x^n$, its average is called the nth moment of *X*. The first moment of *X* is its average, shown above. The second moment of the continuous random variable *x* is

$$\overline{X^2} = \int_{-\infty}^{\infty} x^2 \cdot P(x)dx$$

(10-8)

In the case where the random variable has a nonzero average, *a*, a more useful form of the second moment is the second-order moment about a point *a*, also called the second-order central moment, defined as

$$Var(X) = \int_{-\infty}^{\infty} (x-a)^2 \cdot P(x)dx$$

(10-9)

The second-order central moment of *X* is called the variance of *X*, and its square root is the standard deviation. The standard deviation, commonly represented by the Greek letter sigma (σ), gives a measure of the form factor of a probability density function. Figure 10-4 shows two Gaussian density functions, one with s = 1 and the other with s = 2. Both have the same average, *m*.

The first and second moments are used all the time by electrical engineers when they are talking about voltages and currents, either steady-state or random. The first moment is the DC level and the second moment is proportional to the power. The variance of a voltage across a unit resistance is its AC power, and the standard deviation is the RMS value of a current or voltage about its DC level.

10.2 Information Theory

In 1948 C. E. Shannon published his Mathematical Theory of Communication, which was a tremendous breakthrough in the understanding of the possibilities of reliable communication in the presence of interference. A whole new field of study opened up, that of *information theory*, which deals with three basic concepts: the measure of information, the capacity

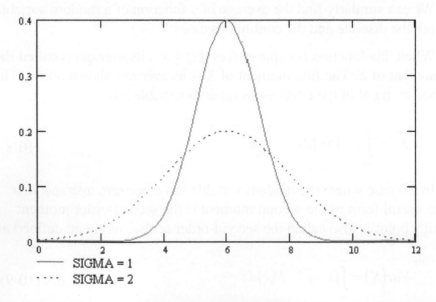

Figure 10-4: Gaussian Density function with Different Standard Deviations

of a communication channel to transfer information, and the use of coding as the means of approaching error-free communication at rates that approach this capacity. Much of the significance of Shannon's work can be realized by considering this statement, which sums up what is called the fundamental theorem of information theory:

It is possible to transmit information through a noisy communication channel at any rate up to the channel capacity with an arbitrarily small probability of error.

The converse of this statement has also been proven:

It is not possible to achieve reliable communication through a channel at a rate higher than the channel capacity.

The key to reliable communication in the presence of noise is *coding*. We will look at an example of error correction coding after a brief review of the basics of information theory.

Uncertainty, Entropy, and Information

Engineers generally talk about *bit* rate as the number of binary symbols that are transmitted per unit time. In information theory, the bit has a different, deeper meaning. Imagine the transmission of an endless stream of digital ones where each one has a duration of one millisecond. Is the rate of communication 1000 bits/second? According to information theory, the rate is zero. A stream of ones, or any other repetitive pattern of symbols reveals nothing to the receiver, and even in the presence of noise and interference, the "message" is always detected with 100 percent certainty. The more uncertain we are about what is being transmitted, the more information we get by correctly receiving it. In information theory, the term "bit" is a unit used to measure a quantity of information or uncertainty.

Information theory defines mathematically the uncertainty of a message or a symbol. Let's say we want to send a message using a sequence of three binary digits. We know that we can send up to eight different messages using this sequence (2^3). Each message has a particular probability of being sent. An example of this situation is a room with 8 patients in a hospital that has a nurse call system. When one of the patients presses a button next to his or her bed, a three-digit message is sent to the nurse's desk where it rings a bell and causes a display to show a number representing that patient. Depending on their condition, some patients may use the call system more than others. We present the probability of a patient pressing the nurse call button in the following table:

Patient's Name	Patient's Number	Probability
John	1	.1
Mary	2	.5
Jane	3	.2
Mike	4	.05
Pete	5	.05
Sue	6	.03
Tom	7	.01
Elaine	8	.06

The total probability of one of them having pressed the call button when the bell rings is, of course, one.

When the nurse hears the bell, she won't be surprised to see the number 2 on the display, since Mary requires assistance more than any of the other patients. The display of "7" though will be quite unexpected, since Tom rarely resorts to calling the nurse. The message indicating that Tom pressed his assistance button gives more information than the one triggered by Mary.

If we label the probability of an event, such as a patient pressing the assistance button, by p_i, we quantify the self-information of the event i by

$$I(i) = \log_2(1/p_i) \tag{10-10}$$

With i referring to the number of the patient, we find that the self-information of Mary's signal is $I(2) = \log_2(2) = 1$ bit, and Tom's signal self-information is $I(7) = \log_2(100) = 6.64$ bits. The unit of self-information is the *bit* when the log is taken to the base 2. The self-information of the other patients' signals can be found similarly.

More important than the individual self-information of each of the signals is the average self-information of all the signals. This quantity is commonly labeled *H* and is called the *uncertainty* of the message, or its *entropy*. The latter term was taken from a value in thermodynamics with similar properties. Taking the statistical average of the self information, we get

$$H = \sum_i p_i I(i)$$

(10-11)

With *i* ranging from 1 to 8 and using the probabilities in the table above, we get for our present example

$H = .1(3.322)+.5(1)+.2(2.322)+.05(2)(4.322)+$
$.03(5.059)+.01(6.644)+.06(4.059)$

$H = 2.191$ bits/message

Now let's assume that all of the patients are equally likely to call the nurse. The self-information of each signal $I(i)$ in this case is $\log_2(8) = 3$ bits. The entropy is

$H_m = 8 \times 1/8 \times 3 = 3$ bits/message

It turns out that this is the maximum possible entropy of the message—the case where the probability of each of the signals is equal.

Let's stretch our example of the nurse call system to an analogy with a continuous stream of binary digits. We'll assume the patients press their call buttons one after another at a constant rate.

In the case where the signal probabilities are distributed as in the table, the entropy per digit is $H/3 = .73$. If each patient pressed his button with equal probability, the entropy per digit would be the maximum of $H_m/3 = 1$. So with the different probabilities as listed in the table, the system communicates only 73% of the information that is possible to send over the communication channel per unit time. Using coding, discussed below, we can match a source to a channel to approach the channel's capability, or capacity, to any extent that we want it to.

Up to now we have been talking about the entropy, or uncertainty, of a source, and have assumed that what is sent is what is received. At least of equal importance is to measure the entropy, and the information, involved with messages sent and received over a noisy channel. Because of the noise, the probabilities of the received messages might not be the same as those of the source messages, because the receiver will make some wrong decisions as to the identity of the source. We saw above that entropy is a function of probabilities, and in the case of communication over a noisy channel, several sets of probability functions can be defined.

On a discrete, memoryless channel (memoryless because the noise affecting one digit is independent of the noise affecting any other digit) we can present the effect of the noise as a matrix of conditional probabilities. Assume a source transmits symbols having one of four letters $x1$, $x2$, $x3$, and $x4$. X is a random variable expressing the transmitted symbol and Y is a random variable for the received symbol. If a symbol $x1$ happens to be sent, the receiver may interpret it as $y1$, $y2$, $y3$, or $y4$, depending on the effect of the random noise at the moment the symbol is sent. There is a probability of receiving $y1$ when $x1$ is sent, another probability of receiving $y2$ when $x1$ is sent, and so on for a matrix of 16 probabilities as shown below:

$$P(Y|X) = \begin{array}{cccc} p(y1|x1) & p(y2|x1) & p(y3|x1) & p(y4|x1) \\ p(y1|x2) & p(y2|x2) & p(y3|x2) & p(y4|x2) \\ p(y1|x3) & p(y2|x3) & p(y3|x3) & p(y4|x3) \\ p(y1|x4) & p(y2|x4) & p(y3|x4) & p(y4|x4) \end{array} \qquad (10\text{-}12)$$

This example shows a square matrix, which means that the receiver will interpret a signal as being one of those letters that it knows can be transmitted, which is the most common situation. However, in the general case, the receiver may assign a larger or smaller number of letters to the signal so the matrix doesn't have to be square.

The conditional entropy of the output Y when the input X is known is:

$$H(Y|X) = -\sum_i p(x_i) \sum_j p(y_j|x_i) \log p(y_j|x_i) \qquad (10\text{-}13)$$

where the sums are over the number of source letters x_i and received letters y_j, and to get units of bits the log is to the base 2. The expression has a minus sign to make the entropy positive, canceling the sign of the log of a fraction, which is negative.

If we look only at the *Y*s that are received, we get a set of probabilities $p(y1)$ through $p(y4)$, and a corresponding uncertainty $H(Y)$.

Knowing the various probabilities that describe a communication system, expressed through the entropies of the source, the received letters, and the conditional entropy of the channel, we can find the important value of the *mutual information* or *transinformation* of the channel. In terms of the entropies we defined above:

$$I(X;Y) = H(Y) - H(Y|X) \qquad \text{(10-14a)}$$

The information associated with the channel is expressed as the reduction in uncertainty of the received letters given by a knowledge of the statistics of the source and the channel. The mutual information can also be expressed as

$$I(X;Y) = H(X) - H(X|Y) \qquad \text{(10-14b)}$$

which shows the reduction of the uncertainty of the source by the entropy of the channel from the point of view of the receiver. The two expressions of mutual information are equal.

Capacity

In the fundamental theorem of information theory, summarized above, the concept of channel capacity is a key attribute. It is connected strongly to the mutual information of the channel. In fact, the capacity is the maximum mutual information that is possible for a channel having a given probability matrix:

$$C = \max I(X;Y) \qquad \text{(10-15)}$$

where the maximization is taken over all sets of source probabilities. For a channel that has a symmetric noise characteristic, so that the channel probability matrix is symmetric, the capacity is easily shown to be

$$C = \log_2(m) - h \qquad \text{(10-16)}$$

where m is the number of different letters for each source symbol and $h = H(Y|X)$, a constant independent of the input distribution when the channel noise is symmetric. This is the case for expression (10-12), when each row of the matrix has the same probabilities except in different orders.

For the noiseless channel, as in the example of the nurse call system, the maximum information that could be transferred was $H_m = 3$ bits/message, achieved when the source messages all have the same probability. Taking that example into the frame of the definition of capacity, Eq. (10-16), the number of different messages is the number of letters per symbol, so for $m = 8$ ("messages")

$$C = \log_2 m = \log_2 8 = 3 . \tag{10-16a}$$

This is a logical extension of the more general case presented previously. Here the constant $h = 0$ for the situation when there is no noise.

Another situation of interest for finding capacity on a discrete memoryless channel is when the noise is such that the output Y is independent of the input X, and the conditional entropy $H(Y|X) = H(Y)$. The mutual information of such a channel $I(X;Y) = H(Y) - H(Y|X) = 0$ and the capacity is also zero.

The notion of channel capacity and the fundamental theorem also hold for continuous, "analog" channels, where signal-to-noise ratio (S/N) and bandwidth (B) are the characterizing parameters. The capacity in this case is given by the Hartley-Shannon law:

$$C = B \log_2 (1 + S/N) \text{ bits/second.} \tag{10-17}$$

The extension from a discrete system to a continuous one is easy to conceive of when we consider that a continuous signal can be converted to a discrete one by sampling, where the sampling rate is a function of the signal bandwidth (at least twice the highest frequency component in the signal).

From a glance at (10-17) we easily see that bandwidth can be traded off for signal-to-noise ratio, or vice-versa, while keeping a constant capacity. Actually, it's not quite that simple since the signal-to-noise ratio itself depends on the bandwidth. We can show this relationship by rewriting (10-17) using $n =$ the noise density. Substituting $N = n \cdot B$:

$$C = B \log_2 (1 + S/nB) \text{ bits/second} \tag{10-18}$$

In this expression the tradeoff between bandwidth and signal-to-noise ratio (or transmitter power) is tempered somewhat but it still exists. This doesn't mean that increasing the bandwidth automatically lets us send a

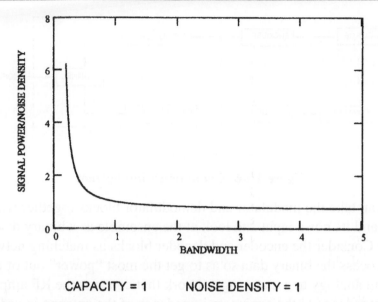

CAPACITY = 1　　　NOISE DENSITY = 1

Figure 10-5: Signal Power vs. Bandwidth at Constant Capacity

higher data rate and keep a low probability of errors. We must use coding to keep the error rate down, but the higher bandwidth facilitates the addition of error-correcting bits in accordance with the coding algorithm.

In radio communication, a common aim is the reduction of signal bandwidth. Shannon-Hartley indicates that we can reduce bandwidth if we increase signal power to keep the capacity constant. This is fine, but Figure 10-5 shows that as the bandwidth is reduced more and more below the capacity, large increases in signal power are needed to maintain that capacity.

Coding

We can represent a communication system conveniently by the block diagram in Figure 10-6. Assume the source outputs binary data, although if it were analog, we could make it binary by passing it through an analog-to-digital converter. The modulator and demodulator act as interfaces between the discrete signal parts of the system and the waveform that passes information over the physical channel—the modulated carrier frequency in a wireless system, for example.

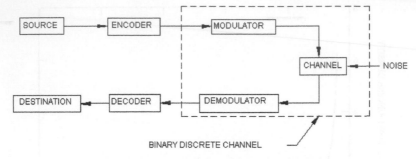

Figure 10-6: Communication System

We can take the modulator and demodulator blocks together with the channel and its noise input and look at the ensemble as a binary discrete channel. Consider the encoder and decoder blocks as matching networks which process the binary data so as to get the most "power" out of the system, in analogy to a matching network that converts the RF amplifier output impedance to the conjugate impedance of the antenna in order to get maximum power transfer. "Power," in the case of the communication link, may be taken to mean the data rate and the equivalent probability of error. A perfect match of the encoder and decoder to the channel gives a rate of information transfer equaling the capacity of the equivalent binary channel with an error rate approaching zero.

Noiseless Coding

When there's no noise in the channel (or relatively little) there is no problem of error rate but coding is still needed to get the maximum rate of information transfer. We saw above that the highest source entropy is obtained when each message has the same probability. If source messages are not equi-probable, then the encoder can determine the message lengths that enter the binary channel so that highly probable messages will be short, and less probable messages will be long. On the average, the channel rate will be obtained.

For example, let's say we have four messages to send over a binary channel and that their rates of occurrence (probabilities) are 1/2, 1/4, 1/8, and 1/8. The following table shows two coding schemes that may be chosen for the messages $m1$, $m2$, $m3$, and $m4$.

Message	Probability	Code A	Code B
*m*1	.5	00	0
*m*2	.25	01	10
*m*3	.125	10	110
*m*4	.125	11	111

One of the possible streams of messages that represents the probabilities is $m1 \cdot m1 \cdot m1 \cdot m1 \cdot m2 \cdot m2 \cdot m3 \cdot m4$. The bit streams that would be produced by each of the two codes is:

CODE A: 0000000001011011

CODE B: 00001010110111

Code A needs 16 binary symbols, or digits, to send the message stream while Code B needs only 14 digits. Thus, using Code B, we can make better use of the binary channel than if we use Code A, since, on the average, it lets us send 14% more messages using the same digit rate.

We can determine the best possible utilization of the channel by calculating the entropy of the messages, which is (from 10-11 above):

$$H = .5 \times \log_2(1/.5) + .25 \times \log_2(1/.25) +$$
$$.25 \times \log_2(1/.125) + .25 \times \log_2(1/.125)$$

$H = 1.75$ digits/message

In the example, we sent 8 messages, which can be achieved using a minimum of $8 \times H = 14$ digits, where each digit carries one bit of information. This we get using Code B.

The capacity of the binary symmetric channel, which sends ones and zeros with equal probability, is (from Eq. 10-16)

$C = \log_2(2) - 0 = 1$ bit/digit.

Each digit on the channel is a bit, so Code B fully utilizes the channel capacity. If each message had equal probability, H would be $\log_2(4) = 2$ bits per message. Then the best code to use would be code A. You can prove it yourself using a message stream with each message occurring an equal number of times and checking how many digits you need for each code. The point is that the chosen code matches the entropy of the source to get the maximum rate of communication.

There are several schemes for determining an optimum code when the message probabilities are known. One of them is Huffman's minimum-redundancy code. This code, like the one in the example above, is easy to decode because each word in the code is distinguishable without a deliminator and decoding is done "on the fly." When input probabilities are not known, such codes are not applicable. An example of a code that works on a bit stream without dependency on source probabilities or knowledge of word lengths is the Lempel-Ziv algorithm, used for file compression. Its basic idea is to look for repeated strings of characters and to specify where a previous version of the string started and how many characters it has.

The various codes used to transfer a given amount of "information" in the shortest time for a given bit rate are often called compression schemes, since they also can be viewed as reducing the number of symbols, or bits, needed to represent this information. One should remember that when the symbols in the input message stream are randomly distributed, coding will not make any difference from the point of view of compression or increasing the message transmission rate.

Error Detection and Correction

A most important use of coding is on noisy channels where you want to reduce the error rate while maintaining a given message transmission rate. We've already learned from Shannon that it's theoretically possible to reduce the error rate to as close to zero as we want, as long as the signal rate is below channel capacity. There's more than one way to look at the advantage of coding for error reduction. For example, if we demand a given maximum error rate in a communication system, we can achieve that goal without coding by increasing transmitter power to raise the signal-to-noise ratio and thereby reduce errors. We can also reduce errors by reducing the transmission rate, allowing us to reduce bandwidth, which will also improve the signal-to-noise ratio (since noise power is proportional to bandwidth). So using coding for error correction lets us either use lower power for a given error rate, or increase transmission speed for the same rate. The reduction in signal-to-noise ratio that can be obtained through the use of coding to achieve a given error rate compared to the signal-to-noise ratio required without coding for the same error rate is defined as the *coding gain*.

A simple and well-known way to increase transmission reliability is to add a parity bit at the end of a block, or sequence, of message bits. The parity bit is chosen to make the number of bits in the message block odd or even. If we send a block of seven message bits, say 0110110, and add an odd parity bit, and the receiver gets the message 01001101, it knows that the message has been corrupted by noise, although it doesn't know which bit is in error. This method of error detection is limited to errors in an odd number of bits, since if, say, two bits were corrupted, the total number of bits is still odd and the errors are not noticed. When the probability of a bit error is relatively low, by far most errors will be in one bit only, so the use of one parity bit may still be useful.

Another simple way to provide error detection is by logically adding together the contents of a number of message blocks and appending the result as an additional block in the message sequence. The receiver performs the same logic summing operation as the transmitter and compares its computation result to the block appended by the transmitter. If the blocks don't match, the receiver knows that one or more bits in the message blocks received are not correct. This method gives a higher probability of detection of errors than does the use of a single parity bit, but if an error is detected, more message blocks must be rejected as having suspected errors.

When highly reliable communication is desired, it's not enough just to know that there is an error in one or more blocks of bits, since the aim of the communication is to receive the whole message reliably. A very common way to achieve this is by an automatic repeat query (ARQ) protocol. After sending a message block, or group of blocks, depending on the method of error detection used, the transmitter stops sending and listens to the channel to get confirmation from the receiver. After the receiver notes that the parity bit or the error detection block indicates no errors, it transmits a short confirmation message to the transmitter. The transmitter waits long enough to receive the confirmation. If the message is confirmed, it transmits the next message. If not, it repeats the previous message.

ARQ can greatly improve communication reliability on a noisy communication link. However, there is a price to pay. First, the receiver must have a transmitter, and the transmitter a receiver. This is OK on a two-way link but may be prohibitive on a one-way link such as exists in most security systems. Second, the transmission rate will be reduced because of the necessity to wait for a response after each short transmission period. If the link is particularly noisy and many retransmissions are required, the repetitions themselves will significantly slow down the communication rate. In spite of its limitations, ARQ is widely used for reliable communications and is particularly effective when combined with a forward error correction scheme as discussed in the next section.

Forward Error Correction (FEC)

Just as adding one odd or even parity bit allows determining if there has been an error in reception, adding additional parity bits can tell the location of the error. For example, if the transmitted message contains 15 bits, including the parity bits, the receiver will need enough information to produce a four-bit word to indicate that there are no errors, (error word 0000), or which bit is in error (one out of 15 error words 0001 through 1111). The receiver might be able to produce this four-bit error detection and correction word from four parity bits (in any case, no less than four parity bits), and then we could send 11 message information bits plus four parity bits and get a capability of detecting and correcting an error in any one bit.

R.W. Hamming devised a relatively simple method of determining parity bits for correcting single-bit errors. If n is the total number of message bits in a sequence to be transmitted, and k is the number of parity bits, the relationship between these numbers permitting correction of one digit is

$$2^k \geq n + 1 \tag{10-19}$$

If we represent m as the number of information bits ($n = m + k$) we can find n and k for several values of m in the following table.

m	4	8	11	26	57
k	3	4	4	5	6
n	7	12	15	31	63

As an example, one possible set of rules for finding the four parity bits for a total message length of 12 bits (8 information bits), derived from Hamming's method, is as follows.

We let $x1$, $x2$, $x3$, and so on, represent the position of the bits in the 12-bit word. Bits $x1$, $x2$, $x4$, and $x8$ are chosen in the transmitter to give even parity when the bits are summed up using binary arithmetic as shown in the four equations below (in binary arithmetic, $0 + 0 = 0$, $0 + 1 = 1$, $1 + 0 = 1$, $1 + 1 = 0$). Let $s1$ through $s4$ represent the results of the equations.

$$x1 + x3 + x5 + x7 + x9 = 0 \qquad\qquad s1$$

$$x2 + x3 + x6 + x7 + x10 + x11 = 0 \qquad\qquad s2$$

$$x4 + x5 + x6 + x7 + x12 = 0 \qquad\qquad s3$$

$$x8 + x9 + x10 + x11 + x12 = 0 \qquad\qquad s4$$

In this scheme, $x1$, $x2$, $x4$, and $x8$ designate the location of the parity bits. If we number the information bits appearing, say, in a byte-wide register of a microcomputer, as $m1$ through $m8$, and we label the parity bits $p1$ through $p4$, then the transmitted 12-bit code word would look like this:

$$p1 \cdot p2 \cdot m1 \cdot p3 \cdot m2 \cdot m3 \cdot m4 \cdot p4 \cdot m5 \cdot m6 \cdot m7 \cdot m8$$

Those parity bits are calculated by the transmitter for even parity as shown in the four equations.

When the receiver receives a code word, it computes the four equations and produces what is called a *syndrome*—a four-bit word composed of $s4 \cdot s3 \cdot s2 \cdot s1$. $s1$ is 0 if the first equation = 0 and 1 otherwise, and so on for $s2$, $s3$, and $s4$. If the syndrome word is 0000, there are no single-bit errors. If there is a single-bit error, the value of the syndrome points to the location of the error in the received code word, and complementing that bit performs the correction. For example, if bit 5 is received in error, the resulting syndrome is 0101, or 5 in decimal notation (you should check this by calculating the equations).

Note that different systems could be used to label the code bits. For example, the four parity bits could be appended after the message bits. In this case a lookup table might be necessary in order to show the correspondence between the syndrome word and the position of the error in the code word.

The one-bit correcting code just described gives an order of magnitude improvement of message error rate compared to transmission of information bytes without parity digits, when the probability of a bit error on the channel is 10^{-2}. However, one consequence of using an error-correcting code is that, if message throughput is to be maintained, a faster bit rate on the channel is required, which entails reduced signal-to-noise ratio because of the wider bandwidth needed for transmission. This rate, for the above example, is 12/8 times the uncoded rate, an increase of 50%. Even so, there is still an advantage to coding, particularly when coding is applied to larger word blocks.

Error probabilities are usually calculated on the basis of independence of the noise from bit to bit, but on a real channel, this is not likely to be the case. Noise and interference tend to occur in bursts, so several adjacent bits may be corrupted. One way to counter the noise bursts is by *interleaving* blocks of code words. For simplicity of explanation, let's say we are using four-bit code words. Interleaving the code words means that after forming each word with its parity bits in the encoder, the transmitter sends the first bit of the first word, then the first bit of the second word, and so on. The order of transmitted bits is best shown as a matrix, as in the following table, where the *a*s are the bits of the first word, and *b*, *c*, and *d* represent the bits of the second, third, and fourth words.

a1	*a2*	*a3*	*a4*
b1	*b2*	*b3*	*b4*
c1	*c2*	*c3*	*c4*
d1	*d2*	*d3*	*d4*

The order of transmission of bits is $a1 \cdot b1 \cdot c1 \cdot d1 \cdot a2 \cdot b2 .... d3 \cdot a4 \cdot b4 \cdot c4 \cdot d4$. In the receiver, the interleaving is decomposed, putting the bits back in their original order, after which the receiver decoder can proceed to perform error detection and correction.

The result of the interleaving is that up to four consecutive bit errors can occur on reception and the decoder can still correct them using a one-bit error correcting scheme. A disadvantage is that there is an additional delay of three word durations until decoded words start appearing at the destination.

Forward error correction schemes which deal with more than one error per block are much more complicated, and more effective, than the Hamming code described here. An important class of codes which are not based on blocks of bits but rather add parity bits based on logic operations on a small number of previous bits is called convolutional codes. In general, coding is a very important aspect of radio communication and its use and continued development is a prime reason why radio communication can continue to replace wires while making more effective use of the available spectrum and limited transmitter power.

10.3 Summary

In this chapter we reviewed the basics of probability and coding. Sets of probabilities were applied both to the different signals generated by the source, as well as to the characteristics of a communication channel corrupted by noise. We saw that associated with a noisy communication channel is a maximum data rate called the capacity of the channel.

Through the use of coding, which involves inserting a redundancy in the transmitted data, it's possible to approach error-free communication up to the capacity of the channel.

Although effective and efficient coding algorithms are quite complicated, the great advances in integrated circuit miniaturization and logic circuit speed in recent years have allowed incorporating error-correction coding in a wide range of wireless communications applications. Many of these applications are in products that are mass produced and relatively low cost—among them cellular telephones and wireless local area networks. This trend will certainly continue and will give even the most simple short-range wireless products "wired" characteristics from the point of view of communication reliability.

This chapter has discussed matters which are not necessarily limited to radio communication, and where they do apply to radio communication, not necessarily to short-range radio. However, anyone concerned with getting the most out of a short-range radio communication link should have some knowledge of information theory and coding. The continuing advances in the integration of complex logic circuits and digital signal processing are bringing small, battery-powered wireless devices closer to the theoretical bounds of error-free communication predicted by Shannon fifty years ago.

CHAPTER 11

Applications and Technologies

An important factor in the widespread penetration of short-range devices into the office and the home is the basing of the most popular applications on industry standards. In this chapter, we take a look at some of these standards and the applications that have emerged from them. Those covered pertain to HomeRF, Wi-Fi, HIPERLAN/2, Bluetooth, and Zigbee. In order to be successful, a standard has to be built so that it can keep abreast of rapid technological advancements by accommodating modifications that don't obsolete earlier devices that were developed to the original version. A case in point is the competition between the WLAN (wireless local area network) standard that was developed by the HomeRF Working Group based on the SWAP (shared wireless access protocol) specification, and IEEE specification 802.11, commonly known as Wi-Fi. The former used frequency-hopping spread-spectrum exclusively, and although some increase of data rate was provided for beyond the original 1 and 2 Mbps, it couldn't keep up with Wi-Fi, which incorporated new bandwidth efficient modulation methods to increase data rates 50-fold while maintaining compatibility with first generation DSSS terminals. Other reasons why HomeRF lost out to Wi-Fi are given below.

Many of the new wireless short-range systems are designed for operation on the 2.4 GHz ISM band, available for license-free operation in North America and Europe, as well as virtually all other regions in the world. Most systems have provisions for handling errors due to interference, but when the density of deployment of one or more systems is high, throughput, voice intelligibility, or quality of service in general is bound to suffer. We will look at some aspects of this problem and methods for solving it in relation to Bluetooth and Wi-Fi.

A relatively new approach to short-range communications with unique technological characteristics is ultra-wideband (UWB) signal generation and detection. UWB promises to add applications and users to short-range communication without impinging on present spectrum use. Additionally, it has other attributes including range finding and high power efficiency that are derived from its basic principles of operation. We present the main features of UWB communication and an introduction to how it works.

11.1 Wireless Local Area Networks (WLAN)

One of the hottest applications of short-range radio communication is wireless local area networks. While the advantage of a wireless versus wired LAN is obvious, the early versions of WLAN had considerably inferior data rates so conversion to wireless was often not worthwhile, particularly when portability is not an issue. However, advanced modulation techniques have allowed wireless throughputs to approach and even exceed those of wired networks, and the popularity of highly portable laptop and handheld computers, along with the decrease in device prices, have made computer networking a common occurrence in multi-computer offices and homes.

There are still three prime disadvantages to wireless networks as compared to wired: range limitation, susceptibility to electromagnetic interference, and security. Direct links may be expected to perform at a top range of 50 to 100 meters depending on frequency band and surroundings. Longer distances and obstacles will reduce data throughput. Greater distances between network participants are achieved by installing additional access points to bridge remote network nodes. Reception of radio signals may be interfered with by other services operating on the same frequency band and in the same vicinity. Wireless transmissions are subject to eavesdropping, and a standardized security implementation in Wi-Fi called WEP (wired equivalent privacy), has been found to be breachable with relative ease by persistent and knowledgeable hackers. More sophisticated encryption techniques can be incorporated, although they may be accompanied by reduction of convenience in setting up connections and possibly in performance.

Various systems of implementation are used in wireless networks. They may be based on an industrial standard, which allows compatibility between devices by different manufacturers, or a proprietary design. The latter would primarily be used in a special purpose network, such as in an industrial application where all devices are made by the same manufacturer and where performance may be improved without the limitations and compromises inherent in a widespread standard.

The HomeRF Working Group

The HomeRF Working Group was established by prominent computer and wireless companies that joined together to establish an open industry specification for wireless digital communication between personal computers and consumer electronic devices anywhere in and around the home. It developed the SWAP specification—Shared Wireless Access Protocol, whose major application was setting up a wireless home network that connects one or more computers with peripherals for the purposes of sharing files, modems, printers, and other electronic devices, including telephones. In addition to acting as a transparent wire replacement medium, it also permitted integration of portable peripherals into a computer network. The originators expected their system to be accepted in the growing number of homes that have two or more personal computers.

Following are the main system technical parameters:

- Frequency-hopping network: 50 hops per second

- Frequency range: 2.4 GHz ISM band

- Transmitter power: 100 milliwatt

- Data rate: 1 Mbps using 2FSK modulation
  2 Mbps using 4FSK modulation

- Range: Covers typical home and yard

- Supported stations: Up to 127 devices per network

- Voice connections: Up to 6 full-duplex conversations

- Data security: Blowfish encryption algorithm
  (over 1 trillion codes)

- Data compression: LZRW3-A (Lempel-Ziv) algorithm

- 48-bit network ID: Enables concurrent operation of
  multiple co-located networks

The HomeRF Working Group ceased activity early in 2003. Several reasons may be cited for its demise. Reduction in prices of its biggest competitor, Wi-Fi, all but eliminated the advantage HomeRF had for home networks—low cost. Incompatibility with Wi-Fi was a liability, since people who used their Wi-Fi equipped laptop computer in the office also needed to use it at home, and a changeover to another terminal accessory after work hours was not an option. If there were some technical advantages to HomeRF, support of voice and connections between peripherals for example, they are becoming insignificant with the development of voice interfaces for Wi-Fi and the introduction of Bluetooth.

Wi-Fi

Wi-Fi is the generic name for all devices based on the IEEE specification 802.11 and its derivatives. It is promoted by the Wi-Fi Alliance that also certifies devices to ensure their interoperability. The original specification is being continually updated by IEEE working groups to incorporate technical improvements and feature enhancements that are agreed upon by a wide representation of potential users and industry representatives. 802.11 is the predominant industrial standard for WLAN and products adhering to it are acceptable for marketing all over the world.

802.11 covers the data link layer of lower-level software, the physical layer hardware definitions, and the interfaces between them. The connection between application software and the wireless hardware is the MAC (medium access control). The basic specification defines three types of wireless communication techniques: DSSS (direct sequence spread spectrum), FSSS (frequency-hopping spread spectrum) and IR (infra-red). The specification is built so that the upper application software doesn't have to know what wireless technique is being used—the MAC interface firmware takes care of that. In fact, application software doesn't have to know that a wireless connection is being used at all and mixed wired and wireless links can coexist in the same network.

Wireless communication according to 802.11 is conducted on the 2.400 to 2.4835 GHz frequency band that is authorized for unlicensed equipment operation in the United States and Canada and most European and other countries. A few countries allow unlicensed use in only a portion of this band. A supplement to the original document, 802.11b, adds increased data rates and other features while retaining compatibility with equipment using the DSSS physical layer of the basic specification. Supplement 802.11a specifies considerably higher rate operation in bands of frequencies between 5.2 and 5.8 GHz. These data rates were made available on the 2.4 GHz band by 802.11g that has downward compatibility with 802.11b.

Network Architecture

Wi-Fi architecture is very flexible, allowing considerable mobility of stations and transparent integration with wired IEEE networks. The transparency comes about because upper application software layers (see below) are not dependent on the actual physical nature of the communication links between stations. Also, all IEEE LAN stations, wired or wireless, use the same 48-bit addressing scheme so an application only has to reference source and destination addresses and the underlying lower-level protocols will do the rest.

Three Wi-Fi network configurations are shown in Figures 11-1 through 11-3. Figure 11-1 shows two unattached basic service sets (BSS), each with two stations (STA). The BSS is the basic building block of an 802.11 WLAN. A station can make *ad hoc* connections with other stations within its wireless communication range but not with those in another BSS that is outside of this range. In order to interconnect terminals that are not in direct range one with the other, the distributed system shown in Figure 11-2 is needed. Here, terminals that are in range of a station designated as an access point (AP) can communicate with other terminals not in direct range but who are associated with the same or another AP. Two or more such access points communicate between themselves either by a wireless or wired medium, and therefore data exchange between all terminals in the network is supported. The important thing here is that the media connecting the STAs with the APs, and connecting the APs among themselves are totally independent.

A network of arbitrary size and complexity can be maintained through the architecture of the extended service set (ESS), shown in Figure 11-3. Here, STAs have full mobility and may move from one BSS to another while remaining in the network. Figure 11-3 shows another element type—a portal. The portal is a gateway between the WLAN and a wired LAN. It connects the medium over which the APs communicate to the medium of the wired LAN—coaxial cable or twisted pair lines, for example.

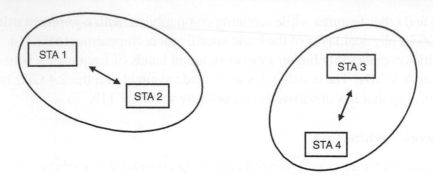

Figure 11-1: Basic Service Set

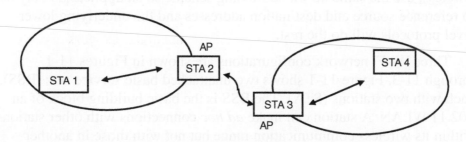

Figure 11-2: Distribution System and Access Points

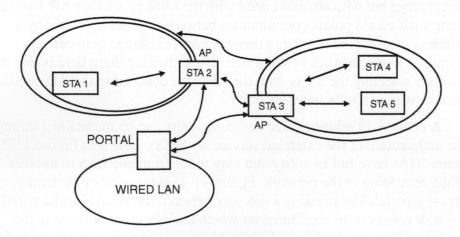

Figure 11-3: Extended Service Set

In addition to the functions Wi-Fi provides for distributing data throughout the network, two other important services, although optionally used, are provided. They are authentication and encryption. Authentication is the procedure used to establish the identity of a station as a member of the set of stations authorized to associate with another station. Encryption applies coding to data to prevent an eavesdropper from intercepting it. 802.11 details the implementation of these services in the MAC. Further protection of confidentiality may be provided by higher software layers in the network that are not part of 802.11.

The operational specifics of WLAN are described in IEEE 802.11 in terms of defined protocols between lower-level software layers. In general, networks may be described by the communication of data and control between adjacent layers of the Open System Interconnection Reference Model (OSI/RM), shown in Figure 11-4, or the peer-to-peer communication between like layers of two or more terminals in the network. The bottom layer, physical, represents the hardware connection with the

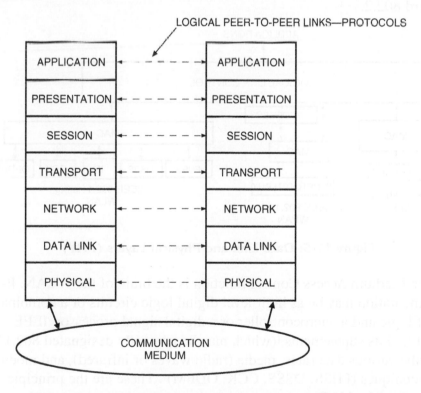

Figure 11-4: Open System Interconnection Reference Model

transmission medium that connects the terminals of the network—cable modem, radio transceiver and antenna, infrared transceiver, or power line transceiver, for example. The software of the upper layers is wholly independent of the transmission medium and in principle may be used unchanged no matter what the nature of the medium and the physical connection to it. IEEE 802.11 is concerned only with the two lowest layers, physical and data link.

IEEE 802.11 prescribes the protocols between the MAC sublayer of the data layer and the physical layer, as well as the electrical specifications of the physical layer. Figure 11-5 illustrates the relationship between the physical and MAC layers of several types of networks with upper-layer application software interfaced through a commonly defined logical link control (LLC) layer. The LLC is common to all IEEE local area networks and is independent of the transmission medium or medium access method. Thus, its protocol is the same for wired local area networks and the various types of wireless networks. It is described in specification ANSI/IEEE standard 802.2.

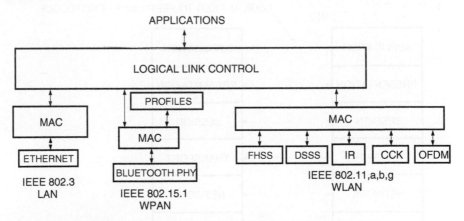

Figure 11-5: Data Link and Physical Layers (PHY)

The Medium Access Control function is the brain of the WLAN. Its implementation may be as high-level digital logic circuits or a combination of logic and a microcontroller or a digital signal processor. IEEE 802.11 and its supplements, (which may be generally designated 802.11x), prescribe various data rates, media (radio waves or infrared), and modulation techniques (FHSS, DSSS, CCK, ODFM) . These are the principle functions of the MAC:

■ Frame delimiting and recognition,

■ Addressing of destination stations,

■ Transparent transfer of data, including fragmentation and defragmentation of packets originating in upper layers,

■ Protection against transmission error,

■ Control of access to the physical medium,

■ Security services—authentication and encryption.

An important attribute of any communications network is the method of access to the medium. 802.11 prescribes two possibilities: DCF (distributed coordination function) and PCF (point coordination function).

The fundamental access method in IEEE 802.11 is the DCF, more widely known as CSMA/CA (carrier sense multiple access with collision avoidance). It is based on a procedure during which a station wanting to transmit may do so only after listening to the channel and determining that it is not busy. If the channel is busy, the station must wait until the channel is idle. In order to minimize the possibility of collisions when more than one station wants to transmit at the same time, each station waits a random time-period, called a back off interval, before transmitting, after the channel goes idle. Figure 11-6 shows how this method works.

The figure shows activity on a channel as it appears to a station that is attempting to transmit. The station may start to transmit if the channel is idle for a period of at least a duration of DIFS (distributed coordination function interframe space) since the end of any other transmission (Section 1 of the figure). However, if the channel is busy, as shown in Section 2 of the figure, it must defer access and enter a back off procedure. The

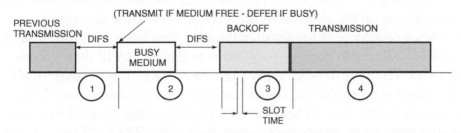

Figure 11-6: CSMA/CA Access Method

station waits until the channel is idle, and then waits an additional period of DIFS. Now it computes a time-period called a back off window that equals a pseudo-random number multiplied by constant called the "slot time." As long as the channel is idle, as it is in Section 3 of the figure, the station may transmit its frame at the end of the back off window, Section 4. During every slot time of the back off window the station senses the channel, and if it is busy, the counter that holds the remaining time of the back off window is frozen until the channel becomes idle and the back off counter resumes counting down.

Actually, the back off procedure is not used for every access of the channel. For example, acknowledgement transmissions and RTS and CTS transmissions, (see below), do not use it. Instead, they access the channel after an interval called SIFS (short interframe space) following the transmission to which they are responding. SIFS is shorter than DIFS, so other stations waiting to transmit cannot interfere since they have to wait a longer time, after the previous transmission, and by then the channel is already occupied.

In waiting for a channel to become idle, a transmission contender doesn't have to listen continuously. When one hears another station access the channel, it can interpret the frame length field that is transmitted on every frame. After taking into account the time of the acknowledgement transmission that replies to a data transmission, the time that the channel will become idle is known even without physically sensing it. This is called a virtual carrier sense mechanism.

The procedure shown in Figure 11-6 may not work well under some circumstances. For example, if several stations are trying to transmit to a single access point, two or more of them may be positioned such that they all are in range of the access point but not of each other. In this case, a station sensing the activity of the channel may not hear another station that is transmitting on the same network. A refinement of the described CSMA/SA procedure is for a station thinking the channel is clear to send a short RTS (request to send) control frame to the AP. It will then wait to receive a CTS (clear to send) reply from the AP, which is in range of all contenders for transmission, before sending its data transmission. If the originating station doesn't hear the CTS it assumes the channel was busy and so it must try to access the channel again. This RTS/CTS procedure is also effective when not all stations on the network have compatible modulation facilities for high rate communication and one station may not be able to detect the transmission length field of another. RTS and CTS transmissions are always sent at a basic rate that is common to all participants in the network.

The PCS is an optional access method that uses a master-slave procedure for polling network members. An AP station assumes the role of master and distributes timing and priority information through beacon management transmissions, thus creating a contention free access method. One use of the PCS is for voice communications, which must use regular time slots and will not work in a random access environment.

Physical Layer

The discussion so far on the services and the organization of the WLAN did not depend on the actual type of wireless connection between the members of the network. 802.11 and its additions specify various bit rates, modulation methods, and operating frequency channels, on two frequency bands, which we discuss in this section.

IEEE 802.11 Basic

The original version of the 802.11 specification prescribes three different air interfaces, each having two data rates. One is infrared and the others are based on frequency-hopping spread spectrum (FHSS) and direct-sequence spread-spectrum, each supporting raw data rates of 1 and 2 Mbps. Below is a short description of the IR and FHSS links, and a more detailed review of DSSS.

Infrared PHY

Infrared communication links have some advantages over radio wave transmissions. They are completely confined within walled enclosures and therefore eavesdropping concerns are greatly relieved, as are problems from external interference. Also, they are not subject to intentional radiation regulations. The IEEE 802.11 IR physical layer is based on diffused infrared links, and the receiving sensor detects radiation reflected off ceilings and walls, making the system independent of line-of-site. The range limit is on the order of 10 meters. Baseband pulse position modulation is used, with a nominal pulse width of 250 nsec. The IR wavelength is between 850 and 950 nM. The 1 Mbps bit rate is achieved by sending symbols representing 4 bits, each consisting of a pulse in one of 16 consecutive 250 nsec slots. This modulation method is called 16-PPM. Optional 4-PPM modulation, with four slots per two-bit symbol, gives a bit rate of 2 Mbps.

Although part of the original IEEE 802.11 specification and having what seems to be useful characteristics for some applications, products based on the infrared physical layer for WLAN have generally not been commercially available. However, point-to-point, very short-range infrared links using the IrDA (Infrared Data Association) standard are very widespread (reputed to be in more than 300 million devices). These links work reliably line-of-site at one meter and are found, for example, in desktop and notebook computers, handheld PC's, printers, cameras and toys. Data rates range from 2400 Bps to 16 Mbps. Bluetooth devices will take over some of the applications but for many cases IrDA imbedding will still have an advantage because of its much higher data rate capability.

FHSS PHY

While overshadowed by the DSSS PHY, acquaintance with the FHSS option in 802.11 is still useful since products based on it may be available. In FHSS WLAN, transmissions occur on carrier frequencies that hop periodically in pseudo-random order over almost the complete span of the 2.4 GHz ISM band. This span in North America and most European countries is 2.400 to 2.4835 GHz, and in these regions there are 79 hopping carrier frequencies from 2.402 to 2.480 GHz. The dwell on each frequency is a system-determined parameter, but the recommended dwell time is 20 msec, giving a hop rate of 50 hops per second. In order for FHSS network stations to be synchronized, they must all use the same pseudo-random sequence of frequencies, and their synthesizers must be in step, that is, they must all be tuned to the same frequency channel at the same time. Synchronization is achieved in 802.11 by sending the essential parameters—dwell time, frequency sequence number, and present channel number—in a frequency parameter set field that is part of a beacon transmission (and other management frames) sent periodically on the channel. A station wishing to join the network can listen to the beacon and synchronize its hop pattern as part of the network association procedure.

The FHSS physical layer uses GFSK (Gaussian frequency shift keying) modulation, and must restrict transmitted bandwidth to 1 MHz at 20 dB down (from peak carrier). This bandwidth holds for both 1 Mbps and 2 Mbps data rates. For 1 Mbps data rate, nominal frequency deviation is ±160 kHz. The data entering the modulator is filtered by a Gaussian (constant phase delay) filter with 3 dB bandwidth of 500 kHz. Receiver sensitivity must be better than –80 dBm for a 3% frame error rate.

In order to keep the same transmitted bandwidth with a data rate of 2 Mbps, four-level frequency shift-keying is employed. Data bits are grouped into symbols of two bits, so each symbol can have one of four levels. Nominal deviations of the four levels are ±72 kHz and ±216 kHz. A 500 kHz Gaussian filter smoothes the four-level 1 Megasymbols per second at the input to the FSK modulator. Minimum required receiver sensitivity is –75 dBm.

Although development of Wi-Fi for significantly increased data rates has been along the lines of DSSS, FHSS does have some advantageous features. Many more independent networks can be collocated with virtually no mutual interference using FHSS than with DSSS. As we will see later, only three independent DSSS networks can be collocated. However, 26 different hopping sequences (North America and Europe) in any of three defined sets can be used in the same area with low probability of collision. Also, the degree of throughput reduction by other 2.4 GHz band users, as well as interference caused to the other users is lower with FHSS. FHSS implementation may at one time also have been less expensive. However, the updated versions of 802.11—specifically 802.11a, 802.11b, and 802.11g—have all based their methods of increasing data rates on the broadband channel characteristics of DSSS in 802.11, while being downward compatible with the 1 and 2 Mbps DSSS modes (except for 802.11a which operates on a different frequency band).

DSSS PHY

The channel characteristics of the direct sequence spread spectrum physical layer in 802.11 are retained in the high data rate updates of the specification. This is natural, since systems based on the newer versions of the specification must retain compatibility with the basic 1 and 2 Mbps physical layer. The channel spectral mask is shown in Figure 11-7, superimposed on the simulated spectrum of a filtered 1 Mbps transmission. It is 22 MHz wide at the –30 dB points. Fourteen channels are allocated in the 2.4 GHz ISM band, whose center frequencies are 5 MHz apart, from 2.412 GHz to 2.484 GHz. The highest channel, number fourteen, is designated for Japan where the allowed band edges are 2.471 GHz and 2.497 GHz. In the US and Canada, the first eleven channels are used. Figure 11-8 shows how channels one, six and eleven may be used by three adjacent independent networks without co-interference. When there are no more than two networks in the same area, they may choose their operating channels to avoid a narrow-band transmission or other interference on the band.

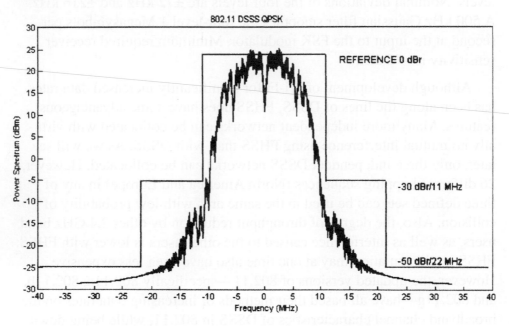

Figure 11-7: 802.11 DSSS Spectral Mask

Figure 11-8: DSSS Non-interfering Channels

In 802.11 DSSS, a pseudo-random bit sequence phase modulates the carrier frequency. In this spreading sequence, bits are called chips. The chip rate is 11 megachips per second (Mcps). Data is applied by phase modulating the spread carrier. There are eleven chips per data symbol. The chosen pseudo-random sequence is a Barker sequence, represented as 1,–1,1,1,–1,1,1,1,–1,–1,–1. Its redeeming property is that it is optimally detected in a receiver by a matched filter or correlation detector. Figure 11-9 is one possible implementation of the modulator. The DSSS PHY specifies two possible data rates—1 and 2 Mbps. The differential encoder takes the data stream and produces two output streams at 1 Mbps that represent changes in data polarity from one symbol to the next. For a data rate of 1 Mbps, differential binary phase shift keying is used. The input data rate of 1 Mbps results in two identical output data streams that represent the changes between consecutive input bits. Differential quadrature phase shift keying handles 2 Mbps of data. Each sequence of two input bits creates four permutations on two outputs. The differential encoder outputs the differences from symbol to symbol on the lines that go to the inputs of the exclusive OR gates shown in Figure 11-9. The outputs on the *I* and *Q* lines are the Barker sequence of 11 Mcps inverted or sent straight through, at a rate of 1 Msps, according to the differentially encoded data at the exclusive OR gate inputs. These outputs are spectrum shifted to the RF carrier frequency (or an intermediate frequency for subsequent up-conversion) in the quadrature modulator.

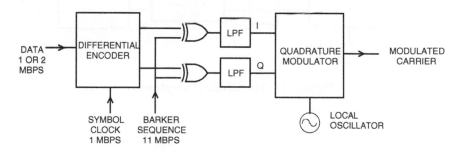

Figure 11-9: DSSS Modulation

Reception of DSSS signals is represented in Figure 11-10. The downconverted *I* and *Q* signals are applied to matched filters or correlation detectors. These circuits correlate the Barker sequence with the input signal and output an analog signal that represents the degree of correlation. The following differential decoder performs the opposite operation of the differential encoder described above and outputs the 1 or 2 Mbps data.

The process of despreading the input signal by correlating it with the stored spreading sequence requires synchronization of the receiver with transmitter timing and frequency. To facilitate this, the transmitted frame starts with a synchronization field (SYNC), shown at the beginning of the physical layer protocol data unit in Figure 11-11. Then a start frame delimiter (SFD) marks out the commencement of the following information bearing fields. All bits in the indicated preamble are transmitted at a rate of 1 Mbps, no matter what the subsequent data rate will be. The signal field specifies the data rate of the following fields in the frame so that the receiver can adjust itself accordingly. The next field, SERVICE, contains all zeros for devices that are only compliant with the basic version of 802.11, but some of its bits are used in devices conforming with updated

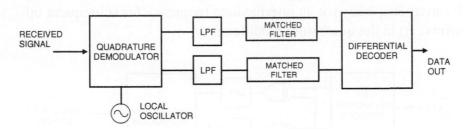

Figure 11-10: DSSS Reception

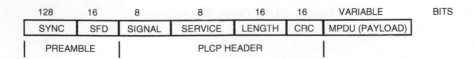

Figure 11-11: DSSS Frame Format

versions. The value of the length field is the length, in microseconds, required to transmit the data-carrying field labeled MPDU (MAC protocol data unit). An error check field, labeled CRC, protects the integrity of the SIGNAL, SERVICE, and LENGTH fields. The last field MPDU (MAC protocol data unit) is the data passed down from the MAC to be sent by the physical layer, or to be passed up to the MAC after reception. All bits in the transmitted frame are pseudo-randomly scrambled to ensure even power distribution over the spectrum. Data is returned to its original form by descrambling in the receiver.

802.11b

The "b" supplement to the original 802.11 specification supports a higher rate physical layer for the 2.4 GHz band. It is this 802.11b version that provided the impetus for Wi-Fi proliferation. With it, data rates of 5.5 Mbps and 11 Mbps are enabled, while retaining downward compatibility with the original 1 and 2 Mbps rates. The slower rates may be used not only for compatibility with devices that aren't capable of the extended rates, but also for fall back when interference or range conditions don't provide the required signal-to-noise ratio for communication using the higher rates.

As previously stated, the increased data rates provided for in 802.11b do not entail a larger channel bandwidth. Also, the narrow-band interference rejection, or jammer resisting qualities of direct sequence spread-spectrum are retained. The classical definition of processing gain for DSSS as being the chip rate divided by the data bandwidth doesn't apply here. In fact, the processing gain requirement that for years was part of the FCC Rules paragraph 15.247 definition of direct sequence spread-spectrum was deleted in an update from August 2002, and at the same time reference to DSSS was replaced by "digital modulation."

The mandatory high-rate modulation method of 802.11b is called complementary code keying (CCK). An optional mode called packet binary convolutional coding (PBCC) is also described in the specification. Although there are similarities in concept, the two modes differ in implementation and performance. First the general principle of high-rate DSSS is presented below, applying to both CCK and PBCC, then the details of CCK are given.

As in the original 802.11, a pseudo-random noise sequence at the rate of 11 Mcps is the basis of high-rate transmission in 802.11b. It is this 11 Mcps modulation that gives the 22 MHz null-to-null bandwidth. However, in contrast to the original specification, the symbol rate when sending data at 5.5 or 11 Mbps is 1.375 Msps. Eight chips per symbol are transmitted instead of eleven chips per symbol as when sending at 1 or 2 Mbps. In "standard" DSSS as used in 802.11, the modulation, BPSK or QPSK, is applied to the group of eleven chips constituting a symbol. The series of eleven chips in the symbol is always the same (the Barker sequence previously defined). In contrast, high-rate DSSS uses a different 8-chip sequence in each symbol, depending on the sequence of data bits that is applied to each symbol. Quadrature modulation is used, and each chip has an I value and a Q value which represent a complex number having a normalized amplitude of one and some angle, α, where α = arctangent (Q/I). α can assume one of four values divided equally around 360 degrees. Since each complex bit has four possible values, there are a total of $4^8 = 65536$ possible 8-bit complex words. For the 11 Mbps data rate, 256 out of these possibilities are actually used—which one being determined by the sequence of 8 data bits applied to a particular symbol. Only 16-chip sequences are needed for the 5.5 Mbps rate, determined by four data bits per symbol. The high-rate algorithm describes the manner in which the 256 code words, or 16 code words, are chosen from the 65536 possibilities. The chosen 256 or 16 complex words have the very desirable property that when correlation detectors are used on the I and Q lines of the received signal, downconverted to baseband, the original 8-bit (11 Mbps rate) or 4-bit (5.5 Mbps rate) sequence can be decoded correctly with high probability even when reception is accompanied by noise and other types of channel distortion.

The concept of CCK modulation and demodulation is shown in Figures 11-12 and 11-13. It's explained below in reference to a data rate of 11 Mbps. The multiplexer of Figure 11-12 takes a block of eight serial data bits, entering at 11 Mbps, and outputs them in parallel, with updates at the symbol rate of 1.375 MHz. The six latest data bits determine 1 out of 64 (2^6) complex code words. Each code word is a sequence of eight complex chips, having phase angles α_1 through α_8 and a magnitude of unity. The first two data bits, d_0 and d_1, determine an angle, α'_8 which, in the code rotator (see Figure 11-12), rotates the whole code word relative to α_8 of

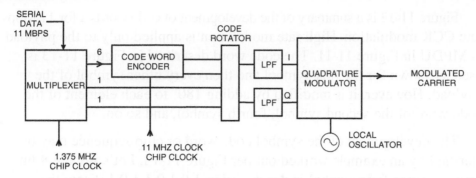

Figure 11-12: High-rate Modulator—11 Mbps

Data symbol: $d_0\ d_1\ d_2\ d_3\ d_4\ d_5\ d_6\ d_7$

Phase Table		
d_i	d_{i+1}	φ
0	0	0^0
1	0	180^0
0	1	90^0
1	1	-90^0

Phase $(d_0, d_1) = \varphi_1$
Phase $(d_2, d_3) = \varphi_2$
Phase $(d_4, d_5) = \varphi_3$
Phase $(d_6, d_7) = \varphi_4$

$\alpha_1 = \varphi_1 + \varphi_2 + \varphi_3 + \varphi_4$
$\alpha_2 = \varphi_1 + \varphi_3 + \varphi_4$
$\alpha_3 = \varphi_1 + \varphi_2 + \varphi_4$
$\alpha_4 = \varphi_1 + \varphi_4 + 180^0$
$\alpha_5 = \varphi_1 + \varphi_2 + \varphi_3$
$\alpha_6 = \varphi_1 + \varphi_3$
$\alpha_7 = \varphi_1 + \varphi_2 + 180^0$
$\alpha_8 = \varphi_1$

$I_i = \cos(\alpha_i)$
$Q_i = \sin(\alpha_i)$
$i = 1...8$

Figure 11-13: Derivation of Code Word

the previous code word. This angle of rotation becomes the absolute angle α_8 of the present code word. The normalized *I* and *Q* outputs of the code rotator, which after filtering are input to a quadrature modulator for up-conversion to the carrier (or intermediate) frequency, are:

$I_i = \cos(\alpha_i),\ Q_i = \sin(\alpha_i)\ \ i = 1 ... 8$.

Figure 11-13 is a summary of the development of code words a for 11 Mbps rate CCK modulation. High rate modulation is applied only to the payload —MPDU in Figure 11-11. The code word described in Figure 11-13 is used as shown for the first symbol and then every other symbol of the payload. However, it is modified by adding 180° to each element of the code word of the second symbol, fourth symbol, and so on.

The development of the symbol code word or chip sequence may be clarified by an example worked out per Figure 11-13. Let's say the 8-bit data sequence for a symbol is $d = d_0 \ldots d_7 = 1\,0\,1\,0\,1\,1\,0\,1$. From the phase table of Figure 11-13 we find the angles φ: $\varphi_1 = 180°$, $\varphi_2 = 180°$, $\varphi_3 = -90°$, $\varphi_4 = 90°$. Now summing up these values to get the angle α_i of each complex chip, then taking the cosine and sine to get I_i and Q_i, we summarize the result in the following table:

i	1	2	3	4	5	6	7	8
α	0	180	90	90	-90	90	180	180
I	1	-1	0	0	0	0	-1	-1
Q	0	0	1	1	-1	1	0	0

The code words for 5.5 Mbps rate CCK modulation are a subset of those for 11 Mbps CCK. In this case, there are four data bits per symbol which determine a total of 16 complex chip sequences. Four 8-element code words (complex chip sequences) are determined using the last two data bits of the symbol, $d2$ and $d3$. The arguments (angles) of these code words are shown in Table 11-1. Bits $d0$ and $d1$ are used to rotate the code words relative to the preceding code word as in 11 Mbps modulation and shown in the phase table of Figure 11-13. Code words are modified by 180° every other symbol, as in 11 Mbps modulation.

Table 11-1: 5.5 Mbps CCK Decoding

$d3,d2$	α_1	α_2	α_3	α_4	α_5	α_6	α_7	α_8
00	90°	0°	90°	180°	90°	0°	-90°	0°
10	-90°	180°	-90°	0°	90°	0°	-90°	0°
01	-90°	0°	-90°	180°	-90°	0°	90°	0°
11	90°	180°	90°	0°	-90°	0°	90°	0°

The concept of CCK decoding for receiving high rate data is shown in Figure 11-14. For the 11 Mbps data rate, a correlation bank decides which of the 64 possible codes best fits each received 8-bit symbol. It also finds the rotation angle of the whole code relative to the previous symbol (one of four values). There are a total of 256 (64 × 4) possibilities and the chosen one is output as serial data. At the 5.5 Mbps rate there are four code words to choose from and after code rotation a total of 16 choices from which to decide on the output data.

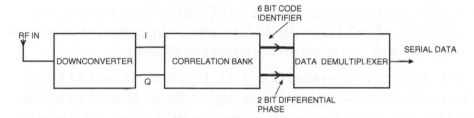

Figure 11-14: CCK Decoding

To maintain compatibility with earlier non high-rate systems, the DSSS frame format shown in Figure 11-11 is retained in 802.11b. The 128-bit preamble and the header are transmitted at 1 Mbps while the payload MPDU can be sent at a high rate of 5.5 or 11 Mbps. The long and slow preamble reduces the throughput and cancels some of the advantage of the high data rates. 802.11b defines an optional short preamble and header which differ from the standard frame by sending a preamble with only 72 bits and transmitting the header at 2 Mbps, for a total overhead of 96 µsec instead of 192 µsec for the long preamble and header. Devices using this option can only communicate with other stations having the same capability.

Use of higher data rates entails some loss of sensitivity and hence range. The minimum specified sensitivity at the 11 Mbps rate is –76 dBm for a frame-error rate of 8% when sending a payload of 1024 bytes, as compared to a sensitivity of –80 dBm for the same frame-error rate and payload length at a data rate of 2 Mbps.

802.11a and OFDM

In the search for ways to communicate at even higher data rates than those applied in 802.11b, a completely different modulation scheme, OFDM (orthogonal frequency division multiplexing) was adopted for 802.11a. It is not DSSS yet it has a channel bandwidth similar to the DSSS systems already discussed. The 802.11*a* supplement is defined for channel frequencies between 5.2 and 5.85 GHz, obviously not compatible with 802.11b signals in the 2.4 GHz band. However, since the channel occupancy characteristics of its modulation are similar to that of DSSS Wi-Fi, the same system was adopted in IEEE 802.11*g* for enabling the high data rates of 802.11a on the 2.4 GHz band, while allowing downward compatibility with transmissions conforming to 802.11b.

802.11a specifies data rates of 6, 9, 12, 18, 24, 36, 48, and 54 Mbit/s. As transmitted data rates go higher and higher, the problem of multipath interference becomes more severe. Reflections in an indoor environment can result in multipath delays on the order of 100 nsec but may be as long as 250 nsec, and a signal with a bit rate of 10 Mbps (period of 100 nsec) can be completely overlapped by its reflection. When there are several reflections, arriving at the receiver at different times, the signal may be mutilated beyond recognition. The OFDM transmission system goes a long way to solving the problem. It does this by sending the data partitioned into symbols whose length in time is several times the expected reflected path length time differences. The individual data bits in a symbol are all sent in parallel on separate subcarrier frequencies within the transmission channel. Thus, by sending many bits during the same time, each on a different frequency, the individual transmitted bit can be lengthened so that it won't be affected by the multipath phenomenon. Actually, the higher bit rates are accommodated by representing a group of data bits by the phase and amplitude of a particular transmitted carrier. A carrier modulated using quadrature phase shift keying (QPSK) can represent two data bits and 64-QAM (quadrature amplitude modulation) can present six data bits as a single data unit on a subcarrier.

Naturally, transmitting many subcarriers on a channel of given width brings up the problem of interference between those subcarriers. There will be no interference between them if all the subcarriers are orthogonal—that is, if the integral of any two different subcarriers over the symbol period is zero. It is easy to show that this condition exists if the frequency difference between adjacent subcarriers is the inverse of the symbol period.

In OFDM, the orthogonal subcarriers are generated mathematically using the inverse Fourier transform (IFT), or rather its discrete equivalent, the inverse discrete Fourier transform (IDFT). The IDFT may be expressed as:

$$x(n) = \frac{1}{N} \sum_{m=0}^{N-1} X(m)[\cos(2\pi mn/N) + j \cdot \sin(2\pi mn/N)]$$

$x(n)$ are complex sample values in the time domain, $n = 0...N-1$, and $X(m)$ are the given complex values, representing magnitude and phase, for each frequency in the frequency domain. The IDFT expression indicates that the time domain signal is the sum of N harmonically related sine and cosine waves each of whose magnitude and phase is given by $X(m)$. We can relate the right side of the expression to absolute frequency by multiplying the arguments $2\pi mn/N$ by f_1/f_s to get

$$x(n) = \frac{1}{N} \sum_{m=0}^{N-1} X(m)[\cos(2\pi mf_1 nt_s) + j \cdot \sin(2\pi mf_1 nt_s)] \tag{11-1}$$

where f_1 is the fundamental subcarrier and the difference between adjacent subcarriers, and t_s is the sample time $1/f_s$. In 802.11a OFDM, the sampling frequency is 20 MHz and $N = 64$, so $f_1 = 312.5$ kHz. Symbol time is $Nt_s = 64/f_s = 3.2$ μsec.

In order to prevent intersymbol interference, 802.11a inserts a guard time of 0.8 μsec in front of each symbol, after the IDFT conversion. During this time, the last 0.8 μsec of the symbol is copied, so the guard time is also called a circular prefix. Thus, the extended symbol time that is transmitted is $3.2 + .8 = 4$ μsec. The guard time is deleted after reception and before reconstruction of the transmitted data.

Although the previous equation, where $N = 64$, indicates 64 possible subcarriers, only 48 are used to carry data, and four more for pilot signals to help the receiver phase lock to the transmitted carriers. The remaining carriers that are those at the outside of the occupied bandwidth, and the DC term ($m = 0$ in Eq.(11-1)), are null. It follows that there are 26

((48 + 4)/2) carriers on each side of the nulled center frequency. Each channel width is 312.5 kHz, so the occupied channels have a total width of 16.5625 (53 × 312.5 kHz) MHz.

For accommodating a wide range of data rates, four modulation schemes are used—BPSK, QPSK, 16-QAM and 64-QAM, requiring 1, 2, 4, and 6 data bits per symbol, respectively. Forward error correction (FEC) coding is employed with OFDM, which entails adding code bits in each symbol. Three coding rates: 1/2, 2/3, and 3/4, indicate the ratio of data bits to the total number of bits per symbol for different degrees of coding performance. FEC permits reconstruction of the correct message in the receiver, even when one or more of the 48 data channels have selective interference that would otherwise result in a lost symbol. Symbol bits are interleaved so that even if adjacent subcarrier bits are demodulated with errors, the error correction procedure will still reproduce the correct symbol. A block diagram of the OFDM transmitter and receiver is shown in Figure 11-15. Blocks FFT and IFFT indicate the fast Fourier transform and its inverse instead of the mathematically equivalent (in terms of results) discrete Fourier transform and inverse discrete Fourier transform (IFDT) that we used above because it is much faster to implement. Table 11-2 lists the modulation type and coding rate used for each data rate, and the total number of bits per OFDM symbol, which includes data bits and code bits.

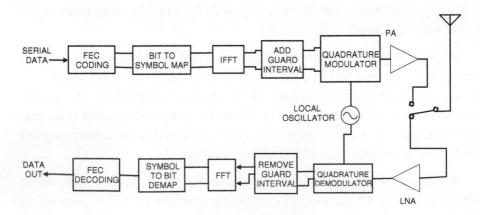

Figure 11-15: OFDM System Block Diagram

Table 11-2: OFDM Characteristics According to Data Rate

Data Rate Mbps	Modulation	Coding Rate	Coded Bits per Subcarrier	Coded Bits per OFDM Symbol	Data Bits per OFDM Symbol
6	BPSK	1/2	1	48	24
9	BPSK	3/4	1	48	36
12	QPSK	1/2	2	96	48
18	QPSK	3/4	2	96	72
24	16-QAM	1/2	4	192	96
36	16-QAM	3/4	4	192	144
48	64-QAM	2/3	6	288	192
54	64-QAM	3/4	6	288	216

The available frequency channels in the 5 GHz band in accordance with FCC paragraphs 15.401–15.407 for unlicensed national information infrastructure (U-NII) devices are shown in Table 11-3. Channel allocations are 5 MHz apart and 20 MHz spacing is needed to prevent co-channel interference. Twelve simultaneous networks can coexist without mutual interference. Power limits are also shown in Table 11-4.

Table 11-3:
Channel Allocations and Maximum Power for 802.11a in United States

Band	Operation Channel Numbers	Channel Center Frequencies (MHz)	Maximum Power with up to 6 dBi antenna gain (mW)
U-NII lower band (5.15–5.25 GHz)	36	5180	40
	40	5200	
	44	5220	
	48	5240	
U-NII middle band (5.25–5.35 GHz)	52	5260	200
	56	5280	
	60	5300	
	64	5320	
U-NII upper band (5.725–5.825 GHz)	149	5745	800
	153	5765	
	157	5785	
	161	5805	

**Table 11-4: HIPERLAN/2 Frequency Channels and Power Levels
(Reference 22, ETSI TS 101 475 V1.3.1 (2001-12))**

Center Frequency (MHz)	*Radiated Power (mean EIRP) (dBm)*
Every 20 MHz from 5180 to 5320	23
Every 20 MHz from 5500 to 5680	30
5700	23

Extension of the data rates of 802.11b to those of 802.11a, but on the 2.4 GHz band is covered in supplement 802.11g. The OFDM physical layer defined for the 5 GHz band is applied essentially unchanged to 2.4 GHz. Equipment complying with 802.11g must also have the lower-rate features and the CCK modulation technique of 802.11b so that it will be downward compatible with existing Wi-Fi systems.

HIPERLAN/2

While 802.11b was designed for compliance with regulations in the European Union and most other regions of the world, 802.11a specifically refers to the regulations of the FCC and the Japanese MPT. ETSI (European Telecommunications Standards Institute) developed a high-speed wireless LAN specification, called HIPERLAN/2 (high performance local area network), which meets the European regulations and in many ways goes beyond the capabilities of 802.11a. HIPERLAN/2 defines a physical layer essentially identical to that of 802.11a, using coded OFDM and the same data rates up to 54 Mbps. However, its second layer software level is very different from the 802.11 MAC and the two systems are not compatible. Built-in features of HIPERLAN/2 that distinguish it from IEEE 802.11a are the following:

- *Quality of service (QOS)*. Time division multiple access/time division duplex (TDMA/TDD) protocol permits multimedia communication.

- *Dynamic frequency selection (DFS)*. Network channels are selected and changed automatically to maintain communication reliability in the presence of interference and path disturbances.

- *Transmit power control (TPC).* Transmission power is automatically regulated to reduce interference to other frequency band users and reduce average power supply consumption.

- *High data security.* Strong authentication and encryption procedures.

All of the above features of HIPERLAN/2 are being dealt with by IEEE task groups for implementation in 802.11. Specifically, the features of DFS and TPC are necessary for conformance of 802.11a to European Union regulations.

Frequency channels and power levels of HIPERLAN/2 are shown in Table 11-4.

11.2 Bluetooth

There are two sources of the Bluetooth specification. One is the Bluetooth Special Interest Group (SIG). The current version at this writing is Version 1.1. It is arranged in two volumes—Core and Profiles. Volume 1, the core, describes the physical, or hardware radio characteristics of Bluetooth, as well as low-level software or firmware which serves as an interface between the radio and higher level specific user software. The profiles in Volume 2 detail protocols and procedures for several widely used applications. The other Bluetooth source specification is IEEE 802.15.1. It is basically a rewriting of the SIG core specification, made to fit the format of IEEE communications specifications in general.

Bluetooth is an example of a wireless personal area network (WPAN), as opposed to a wireless local area network (WLAN). It's based on the creation of *ad hoc*, or temporary, on-the-fly connections between digital devices associated with an individual person and located in the vicinity of around ten meters from him. Bluetooth devices in a network have the function of a master or a slave, and all communication is between a master and one or more slaves, never directly between slaves. The basic Bluetooth network is called a piconet. It has one master and from one to seven slaves. A scatternet is an interrelated network of piconets where any member of a piconet may also belong to an adjacent piconet. Thus, conceptually, a Bluetooth network is infinitely expandable. Figure 11-16 shows a scatternet made up of three piconets. In it, a slave in one piconet is a master in another. A device may be a master in one piconet only.

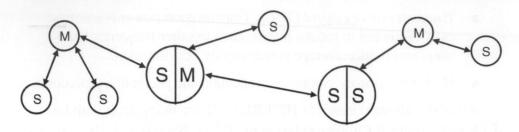

Figure 11-16: Bluetooth Scatternet

The basic RF communication characteristics of Bluetooth are shown in Table 11-5.

Table 11-5: Bluetooth Technical Parameters

CHARACTERISTIC	*VALUE*	*COMMENT*
Frequency Band	2.4 to 2.483 GHz	May differ in some countries
Frequency Hopping Spread Spectrum (FHSS)	79 1-MHz channels from 2402 to 2480 MHz	May differ in some countries
Hop Rate	1600 hops per second	
Channel Bandwidth	1 MHz	20 dB down at edges
Modulation	Gaussian Frequency Shift Keying (GFSK)	
	Filter BT = 0.5	Gaussian Filter bandwidth = 500 kHz
	Nominal modulation index = 0.32	Nominal deviation = 160 kHz
Symbol Rate	1 Mbps	
Transmitter Maximum Power		
Class 1	100 mW	Power control required
Class 2	2.5 mW	Must be at least 0.25 mW
Class 3	1 mW	No minimum specified
Receiver Sensitivity	−70 dBm for BER = 0.1%	

A block diagram of a Bluetooth transceiver is shown in Figure 11-17. It's divided into three basic parts: RF, baseband, and application software. A Bluetooth chip set will usually include the RF and baseband parts, with the application software being contained in the system's computer or controller. The user data stream originates and terminates in the application software. The baseband section manipulates the data and forms frames or data bursts for transmission. It also controls the frequency synthesizer according to the Bluetooth frequency-hopping protocol. The blocks in Figure 11-17 are general and various transmitter and receiver configurations are adopted by different manufacturers. The Gaussian low-pass filter block before the modulator, for example, may be implemented digitally as part of a complex signal *I/Q* modulation unit or it may be a discrete element filter whose output is applied to the frequency control line of a VCO. Similarly, the receiver may be one of several types, as discussed in Chapter 6. If a superheterodyne configuration is chosen, the filter at the output of the downconverter will be a bandpass type. A direct conversion receiver will use low pass filters in complex *I* and *Q* outputs of the downconverter. While different manufacturers employ a variety of methods to implement the Bluetooth radio, all must comply with the same strictly defined Bluetooth specification, and therefore the actual configuration used in a particular chipset should be of little concern to the end user.

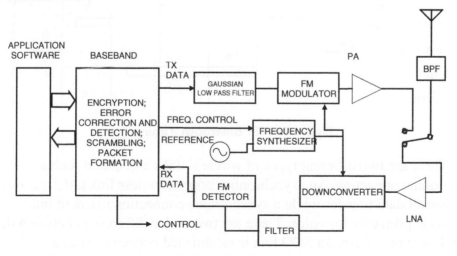

Figure 11-17: Bluetooth Transceiver

The Bluetooth protocol has a fixed-time slot of 625 microseconds, which is the inverse of the hop rate given in Table 11-5. A transmission burst may occur within a duration of one, three, or five consecutive slots on one hop channel. As mentioned, transmissions are always between the piconet master and a slave, or several slaves in the case of a broadcast, or point-to-multipoint transmission. All slaves in the piconet have an internal timer synchronized to the master device timer, and the state of this timer determines the transmission hop frequency of the master and that of the response of a designated slave. Figure 11-18 shows a sequence of transmissions between a master and two slaves. Slots are numbered according to the state, or phase, of the master clock, which is copied to each slave when it joins the piconet. Note that master transmissions take place during even numbered clock phases and slave transmissions during odd numbered phases. Transmission frequency depends on the clock phase, and if a device makes a three or five slot transmission (slave two in the diagram), the intermediate frequencies that would have been used if only single slots were transmitted are omitted (f_4 and f_5 in this case). Note that transmissions do not take up a whole slot. Typically, a single-slot transmission burst lasts 366 microseconds, leaving 259 microseconds for changing the frequency of the synthesizer, phase locked loop settling time, and for switching the transceiver between transmit and receive modes.

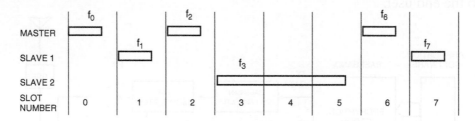

Figure 11-18: Bluetooth Timing

There are two different types of wireless links associated with a Bluetooth connection. An asynchronous connectionless link (ACL) is used for packet data transfer while a synchronous connection oriented link (SCO) is primarily for voice. There are two major differences between the two link types. When an SCO link is established between a master and a slave, transmissions take place on dedicated slots with a constant interval between them. Also, unlike an ACL link, transmitted frames are not repeated in the case of an error in reception. Both of these conditions are necessary because voice is a continuous real-time process whose data rate cannot be randomly varied without affecting intelligibility. On the other hand, packet data transmission can use a handshaking protocol to regulate data accumulation and the instantaneous rate is not usually critical. Thus, for ACL links the master has considerable leeway in proportioning data transfer with the slaves in its network. An ARQ (automatic repeat request) protocol is always used, in addition to optional error correction, to ensure the highest reliability of the data transfer.

Bluetooth was conceived for employment in mobile and portable devices, which are more likely than not to be powered by batteries, so power consumption is an important issue. In addition to achieving low-power consumption due to relatively low transmitting power levels, Bluetooth incorporates power saving features in its communication protocol. Low average power is achieved by reducing the transmission duty cycle, and putting the device in a low-power standby mode for as long a period as possible relative to transmit and receive times while still maintaining the minimum data flow requirements. Bluetooth has three modes for achieving different degrees of power consumption during operation: sniff, hold, and park. Even in the normal active mode, some power saving can be achieved, as described below.

Active Mode

During normal operation, a slave can transmit in a particular time slot only if it is specifically addressed by the master in the proceeding slot. As soon as it sees that its address is not contained in the header of the master's message, it can "go to sleep," or enter a low-power state until it's time for the next master transmission. The master also indicates the length of its transmission (one, three, or five slots) in its message header, so the slave can extend its sleep time during a multiple slot interval.

Sniff Mode

In this mode, sleep time is increased because the slave knows in advance the time interval between slots during which the master may address the slave. If it's not addressed during the agreed slot, it returns to its low-power state for the same period and then wakes up and listens again. When it is addressed, the slave continues listening during subsequent master transmission slots as long as it is addressed, or for an agreed time-out period.

Hold Mode

The master can put a slave in the hold mode when data transfer between them is being suspended for a given period of time. The slave is then free to enter a low-power state, or do something else, like participate in another piconet. It still maintains its membership in the original piconet, however. At the end of the agreed time interval, the slave resynchronizes with the traffic on the piconet and waits for instructions from the master.

Park Mode

Park has the greatest potential for power conservation, but as opposed to hold and sniff, it is not a directly addressable member of the piconet. While it is outside of direct calling, a slave in park mode can continue to be synchronized with the piconet and can rejoin it later, either on its own initiative or that of the master, in a manner that is faster than if it had to join the piconet from scratch. In addition to saving power, park mode can also be considered a way to virtually increase the network's capacity from eight devices to 255, or even more. When entering park mode, a slave gives up its active piconet address and receives an 8-bit parked member address. It goes into low-power mode but wakes up from time to time to listen to the traffic and maintain synchronization. The master sends beacon transmissions periodically to keep the network active. Broadcast transmissions to all parked devices can be used to invite any of them to rejoin the network. Parked units themselves can request re-association with the active network by way of messages sent during an access window that occurs a set time after what is called a "beacon instant." A polling technique is used to prevent collisions.

Packet Format

In addition to the data that originates in the high-level application software, Bluetooth packets contain fields of bits that are created in the baseband hardware or firmware for the purpose of acquisition, addressing, and flow control. Packet bits are also subjected to data whitening (randomization), error-correction coding, and encryption as defined for each particular data type. Figure 11-19 shows the standard packet format.

ACCESS CODE	HEADER	PAYLOAD
72 BITS	54 BITS	0 TO 2745 BITS

Figure 11-19: Bluetooth Packet

The access code is used for synchronization, d-c level compensation, and identification. Each Bluetooth device has a unique address, and it is the address of the device acting as master that is used to identify transmitted packets as belonging to a specific piconet. A 64-bit synchronization word sandwiched between a four-bit header and four-bit trailer, which provide d-c compensation, is based on the master's address. This word has excellent correlation properties so when it is received by any of the piconet members it provides synchronization and positive identification that the packet of which it is a part belongs to their network. All message packets sent by members of the piconet use the same access code.

The header contains six fields with link control information. First, it has a three-bit active member address which identifies to which of the up to seven slaves a master's message is destined. An all zero address signifies a broadcast message to all slaves in the piconet. The next field has four bits that define the type of packet being sent. It specifies, for example, whether one, three, or five slots are occupied, and the level of error correction applied. The remaining fields involve flow control (handshaking), error detection and sequencing. Since the header has prime importance in the packet, it is endowed with forward-error correction having a redundancy of times three.

Following the header in the packet is the payload, which contains the actual application or control data being transferred between Bluetooth devices. The contents of the payload field depend on whether the link is an ACL or SCO. The payload of ACL links has a payload header field that specifies the number of data bytes and also has a handshaking bit for data-buffering control. A CRC (cyclic redundancy check) field is included for data integrity. As stated above, SCO links don't retransmit packets so they don't include a CRC. They don't need a header either because the SCO payload has a constant length.

The previous packet description covers packets used to transfer user data, but other types of packets exist. For example, the minimum length packet contains only the access code, without the four-bit trailer, for a total of 68 bits. It's used in the inquiry and paging procedures for initial frequency-hopping synchronization. There are also NULL and POLL packets that have an access code and header, but no payload. They're sent when slaves are being polled to maintain synchronization or confirm packet reception (in the case of NULL) in the piconet but there is no data to be transferred.

Error Correction and Encryption

The use of forward error correction (FEC) improves throughput on noisy channels because it reduces the number of bad packets that have to be retransmitted. In the case of SCO links that don't use retransmission, FEC can improve voice quality. However, error correction involves bit redundancy so using it on relatively noiseless links will decrease throughput. Therefore, the application decides whether to use FEC or not.

As already mentioned, there are various types of packets, and the packet type defines whether or not FEC is used. The most redundant FEC method is always used in the packet header, and for the payload in one type of SCO packet. It simply repeats each bit three times, allowing the receiver to decide on the basis of majority rule what data bit to assign to each group of incoming bits.

The other FEC method, applied in certain type ACL and SCO packets, uses what's called a (15,10) shortened Hamming code. For every ten data bits, five parity bits are generated. Since out of every 15 transmitted bits only ten are retrieved, the data rate is only two-thirds what it would be without coding. This code can correct all single errors and detect all double errors in each 15-bit code word.

Wireless communication is susceptible to eavesdropping so Bluetooth incorporates optional security measures for authentication and encryption. Authentication is a procedure for verifying that received messages are actually from the party we expect them to be and not from an outsider who is inserting false messages. Encryption prevents an eavesdropper from understanding intercepted communications, since only the intended recipient can decipher them. Both authentication and implementation routines are implemented in the same way. They involve the creation of secret keys that are generated from the unique Bluetooth device address, a PIN (personal identification number) code, and a random number derived from a random or pseudo-random process in the Bluetooth unit. Random numbers and keys are changed frequently. The length of a key is a measure of the difficulty of cracking a code. Authentication in Bluetooth uses a 128-bit key, but the key size for encryption is variable and may range from 8 to 128 bits.

Inquiry and Paging

A distinguishing feature of Bluetooth is its *ad hoc* protocol and connections are often required between devices that have no previous knowledge of their nature or address. Also, Bluetooth networks are highly volatile, in comparison to WLAN for example, and connections are made and dissolved with relative frequency. To make a new connection, the initiator —the master—must know the address of the new slave, and the slave has to synchronize its clock to the master's in order to align transmit and receive channel hop-timing and frequencies. The inquiry and paging procedures are used to create the connections between devices in the piconet.

By use of the inquiry procedure, a connection initiator creates a list of Bluetooth devices within range. Later, desired units can be summoned into the piconet of which the initiator is master by means of the paging routine.

As mentioned previously, the access code contains a synchronization word based on the address of the master. During inquiry, the access code is a general inquiry access code (GIAC) formed from a reserved address for this purpose. Dedicated inquiry access codes (DIAC) can also be used when the initiator is looking only for certain types of devices. Now a potential slave can lock on to the master, provided it is receiving during the master's transmission time and on the transmission frequency. To facilitate this match-up, the inquiry procedure uses a special frequency hop routine and timing. Only 32 frequency channels are used and the initiator transmits two burst hops per standard time slot instead of one. On the slot following the transmission inquiry bursts, the initiator listens for a response from a potential slave on two consecutive receive channels whose frequencies are dependent on the previously transmitted frequencies.

When a device is making itself available for an inquiring master, it remains tuned to a single frequency for a period of 1.28 seconds and at a defined interval and duration scans the channel for a transmission. At the end of the 1.28-second period, it changes to another channel frequency. Since the master is sending bursts over the whole inquiry frequency range at a fast rate—two bursts per 1250 microsecond interval—there's a high probability the scanning device will catch at least one of the transmissions while it remains on a single frequency. If that channel happens to be blocked by interference, then the slave will receive a transmission after one of its subsequent frequency changes. When the slave does hear a signal, it responds during the next slot with a special packet called FHS (frequency hop synchronization) in which is contained the slave's Bluetooth address and state of its internal clock register. The master does not respond but notes the slave's particulars and continues inquiries until it has listed the available devices in its range. The protocol has provisions for avoiding collisions from more than one scanning device that may have detected a master on the same frequency and at the same time.

The master makes the actual connection with a new device appearing in its inquiry list using the page routine. The paging procedure is quite similar to that of the inquiry. However, now the master knows the paged device's address and can use it to form the synchronization word in its access code. The designated slave does its page scan while expecting the access code derived from its own address. The hopping sequence is different during paging than during inquiry, but the master's transmission bursts and the slave's scanning routine are very similar.

A diagram of the page state transmissions is given in Figure 11-20. When the slave detects a transmission from the master (Step 1), it responds with a burst of access code based on its own Bluetooth address. The master then transmits the FHS, giving the slave the access code information (based on the master's address), timing and piconet active member address (between one and seven) needed to participate in the network. The slave acknowledges FHS receipt in Step 4. Steps 5 and 6 show the beginning of the network transmissions which use the normal 79 channel hopping-sequence based on the master's address and timing.

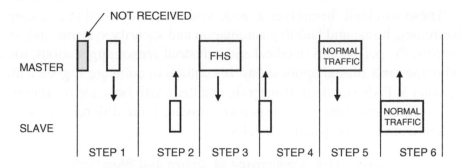

Figure 11-20: Paging Transmissions

11.3 Zigbee

Zigbee is the name of a standards-based wireless network technology that addresses remote monitoring and control applications. Its promotion and development is being handled on two levels. A technical specification for the physical and data link layers, IEEE 802.15.4, was drawn up by a working group of the IEEE as a low data rate WPAN (wireless personal area network). An association of committed companies, the Zigbee Alliance, is defining the network, security, and application layers above the 802.15.4 physical and medium access control layers, and will deal with interoperability certification and testing.

The distinguishing features of Zigbee to which the IEEE standard addresses itself are

- Low data rates—throughput between 10 and 115.2 Kbps

- Low power consumption—several months up to two years on standard primary batteries

- Network topology appropriate for multisensor monitoring and control applications

- Low complexity for low cost and ease of use

- Very high reliability and security

These will lend themselves to wide-scale use embedded in consumer electronics, home and building automation and security systems, industrial controls, PC peripherals, medical and industrial sensor applications, toys and games and similar applications. It's natural to compare Zigbee with the other WPAN standard, Bluetooth, and there will be some overlap in implementations. However, the two systems are quite different, as is evident from the comparison in Table 11-6.

Table 11-6: Comparison of Zigbee and Bluetooth

	Bluetooth	*Zigbee*
Transmission Scheme	FHSS (Frequency Hopping Spread Spectrum)	DSSS (Direct Sequence Spread Spectrum)
Modulation	GFSK (Gaussian Frequency Shift Keying)	QPSK (Quadrature Phase Shift Keying) or BPSK (Binary Phase Shift Keying)
Frequency Band	2.4 GHz	2.4 GHz, 915 MHz, 868 MHz
Raw Data Bit Rate	1 MBPS	250 KBPS, 40 KBPS or 20 KBPS (depends on frequency band)
Power Output	Maximum 100 mW, 2.5 mW, or 1 mW, depending on class	Minimum capability 0.5 mW; maximum as allowed by local regulations
Minimum Sensitivity	–70 dBm for 0.1% BER	-85 dBm (2.4 GHz) or –92 dBm (915/868 MHz) for packet error rate < 1%
Network topology	Master-Slave 8 active nodes	Star or Peer-Peer 255 active nodes

Architecture

The basic architecture of Zigbee is similar to that of other IEEE standards, Wi-Fi and Bluetooth for example, a simplified representation of which is shown in Figure 11-21. On the bottom are the physical layers, showing two alternative options for the RF transceiver functions of the specification. Both of these options are never expected to exist in a single device, and indeed their transmission characteristics—frequencies, data rates, modulation system—are quite different. However, the embedded firmware and software layers above them will be essentially the same no matter what physical layer is applied. Just above the physical layers is the data link layer, consisting of two sublayers: medium access control, or MAC, and the logical link control, LLC. The MAC is responsible for management of the physical layer and among its functions are channel access, keeping track of slot times, and message delivery acknowledgement. The LLC is the interface between the MAC and physical layer and the upper-application software.

Application software is not a part of the IEEE 802.15.4 specification and it is expected that the Zigbee Alliance will prepare profiles, or programming guidelines and requirements for various functional classes in order to assure product interoperability and vendor independence. These profiles will define network formation, security, and application requirements while keeping in mind the basic Zigbee features of low power and high reliability.

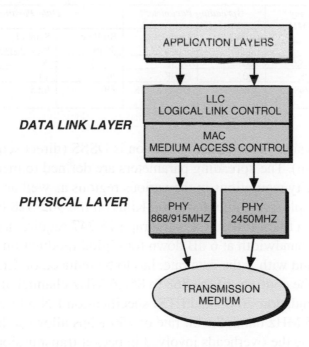

Figure 11-21: Zigbee Architecture

Communication Characteristics

In order to achieve high flexibility of adaptation to the range of applications envisioned for Zigbee, operation is being specified for three unlicensed bands—2.4 GHz, 915 MHz and 868 MHz; the latter two being included in the same physical layer. Those two bands are generally mutually exclusive, their use being determined by geographic location and regional regulations. The following 27 transmitting channels are defined:

Channel Number	Center Frequency Range	Channel Width
0	868.3 MHz	600 kHz
1 to 10	906 to 924 MHz	2 MHz
11 to 27	2405 to 2480	5 MHz

Data rates and modulation types for each of the bands are shown in Table 11-7.

Table 11-7: Data Rates and Modulation

PHY (MHz)	Frequency Band (MHz)	Spreading Parameters		Data Parameters		
		Chip Rate (kcps)	Modulation	Bit Rate (kbps)	Symbol Rate (ksps)	Symbols
868/915	868–868.6	300	BPSK	20	20	Binary
	902–928	600	BPSK	40	40	Binary
2450	2400–2483.5	2000	Offset-QPSK	250	62.5	16-ary Orthogonal

In both physical layers, the modulation is DSSS (direct sequence spread spectrum). The spreading parameters are defined to meet communication authority regulations in the various regions as well as desired data rates. For example, the chip rate of 600 Kbps on the 902-928 band allows the transmission to meet the FCC paragraph 15.247 requirement of minimum 500 kHz bandwidth at 6 dB down for digital modulation. However the chip rate, and with it the data rate, has to be reduced on Channel 0 in order to meet the confines of the 868 to 868.6 MHz channel allowed under ERC recommendation 70-03 and ETSI specification EN 300-220. On the 2400 to 2483.5 MHz band, the bit rate of 250 Kbps allows a throughput, after considering the overheads involved in packet transmissions, to attain 115.2 Kbps, a rate used for some PC peripherals for example.

The spreading modulation used on the 2450 MHz physical layer has similarity in principle to that used on IEEE 802.11b (high-rate Wi-Fi) to increase the data bit rate without raising the chip rate, thereby achieving a desired carrier bandwidth. Sixteen different, almost orthogonal 32-bit long spreading sequences are available for transmission at 2 Mchips/second. Each consecutive sequence of four data bits determines which of the sixteen spreading sequences is sent. On reception, the receiver can identify the spreading sequence and thus decode the data bits. The modulation used, O-QPSK (offset quadrature phase shift keying) with half-sine wave pulse shaping is essentially equivalent to a form of frequency shift keying, MSK (minimum shift keying). It is fairly easy to generate and has a relatively narrow bandwidth for the given chip rate. This latter feature allows a large number of nonoverlapping channels that can be used, with proper upper layer software, on the crowded 2.4 GHz band to avoid interference.

Other physical layer characteristics of Zigbee are output power and receiver sensitivity. The devices must be capable of radiating at least –3 dBm although output may be reduced to the minimum necessary in order to limit interference to other users. Maximum power is determined by the regulatory authorities. While much higher powers are allowed, it may not be practical to transmit over, say 10 dBm, because of absolute limits on spurious radiation and the general objective of low-cost and low-power consumption. Minimum receiver sensitivity for the 868/915 MHz physical layer is specified as –92 dBm and –85 dBm on 2.4 GHz. These limits are for a packet error rate of one percent.

Device Types and Topologies

Two device types, of different complexities, are defined. A full function device (FFD) will be able to implement the full protocol set and can act as a network coordinator. Devices capable of minimal protocol implementation are reduced function devices (RFD). Due to the distinction between device types, networks in which most members require only minimum functionality, such as switches and sensors, can be made significantly less costly and have lower power consumption than if all devices were constrained to have maximum capability.

Flexibility in network configuration is achieved through two topologies—star and peer-to-peer that are depicted in Figure 11-22. A network may have as many as 255 members, one of which is a PAN (personal area network) coordinator. The function of the PAN coordinator, in addition to any specific application it may have, is to initiate, terminate, or route communication around the network. It also provides synchronization services. In a star network, each device communicates directly with the coordinator. The coordinator must be a FFD, and the others can be FFDs or RFDs. Relatively simple applications, like PC peripherals and toys, would typically use the star topology.

In the peer-to-peer topology, any device can communicate with any other device as long as it is in range. RFDs cannot participate, since an RFD can only communicate with a FFD. More complicated structures can be set up as a combination of peer-to-peer groups and star configurations. There is still just one PAN coordinator in the whole network. One example of such a structure is a cluster-tree network shown in Figure 11-23. In this arrangement devices on the network extremities may well be out of radio range of each other, but they can still communicate by relaying messages through the individual clusters.

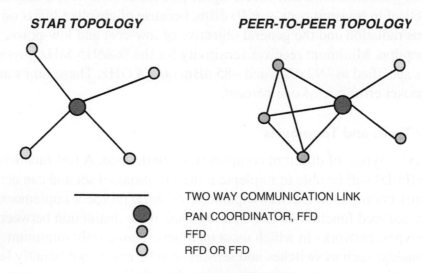

STAR TOPOLOGY   **PEER-TO-PEER TOPOLOGY**

――――――――― TWO WAY COMMUNICATION LINK
● PAN COORDINATOR, FFD
◍ FFD
○ RFD OR FFD

Figure 11-22: Network Topologies

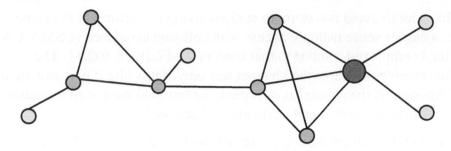

Figure 11-23: Cluster-Tree Topology

Frame Structure, Collision Avoidance, and Reliability

Zigbee frame construction and channel access are similar to those of WLAN 802.11 (Wi-Fi) but are less complex. The transmitted packet has the basic construction shown in Figure 11-24. The purpose of the preamble is to permit acquisition of chip and symbol timing. The PHY header, which is signaled by a delimiter byte, notifies the baseband software in the receiver of the length of the subsequent data. The PSDU (PHY service data unit) is the message that has been passed down through the higher protocol layers. As shown, it can have a maximum of 127 bytes although monitoring and control applications will typically be much shorter. Included in the PSDU are information on the format of the message frame, a sequence number, address information, the data payload itself, and at the end, two bytes that serve as a frame check sequence. Reliability is assured since the receiver performs an independent calculation of this frame check sequence and compares it with the number received. If any bits have been changed by interference or noise, the numbers will not match. Only if a match occurs, the receiving side returns an acknowledgement to the originator of the message. Lacking an acknowledgement, the transmission will be repeated until it is successfully received.

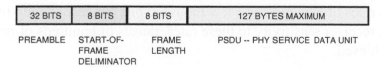

Figure 11-24: Transmission Packet

In order to avoid two or more stations trying to transmit at the same time, a carrier sense multiple access with collision avoidance (CSMA-CA) routine is employed, similar to that used in Wi-Fi, IEEE 802.11. The Zigbee receiver monitors the channel and only if it is idle it may initiate a transmission. If the channel is occupied, the terminal must wait a random back off period before it can again attempt access.

Acknowledgement messages are sent without using the collision avoidance mechanism.

Zigbee Applications

While the promoters of Zigbee aim to cover a very large market for those applications that require relatively low data rates, there will remain applications for which the compromises inherent in a general specification are not acceptable, and producers will continue to develop devices with proprietary specifications and characteristics. However, the open specification and a recognized certification of conformity are an advantage in many situations. For example, a home burglar alarm system would accept wireless sensors produced by different manufacturers, which will facilitate future expansion or allow installers to add sensors of types not available from the original system manufacturer. Use of devices approved according to a recognized standard gives the consumer some security against obsolescence.

Although Zigbee claims to be appropriate for most control applications, it will not fit all of them, and will not necessarily take advantage of all the possibilities of the unlicensed device regulations. Its declared maximum range of some 50 to 75 meters will fall short of the requirements of many systems. Given the meager maximum power allowed, greater range means reduced bandwidth and reduced data rate. In fact, a great many of the applications envisaged by Zigbee can get by very well with data rates of hundreds or a few thousand bits per second, and by matching receiver sensitivity to these rates, ranges of hundreds of meters can be achieved.

One partial answer to the range question is the deployment of the Zigbee network in a cluster-tree configuration, as previously described. Adjacent nodes serve as repeaters so that large areas can be covered, as long as the greatest distance between any two directly communicating nodes does not exceed Zigbee's basic range capability. For example, in a multi-floor building, sensors on the top floor can send alarms to the control box in the basement by passing messages through sensors located on every floor and operating as relay stations.

No doubt that there will be competition between Bluetooth and Zigbee for use in certain applications, but the overall deployment and the reliability of wireless control systems will increase. The proportion of wireless security and automation systems will increase because the new standard will provide a significant boost in reliability, security, and convenience, as compared to most present solutions.

11.4 Conflict and Compatibility

With the steep rise of Bluetooth product sales and the already large and growing use of wireless local area networks, there is considerable concern about mutual interference between Bluetooth-enabled and Wi-Fi devices. Both occupy the 2.4 to 2.4835 GHz unlicensed band and use wideband spread-spectrum modulating techniques. They will most likely be operating concurrently in the same environments, particularly office/commercial but also in the home.

Interference can occur when a terminal of one network transmits on or near the receiving frequency of a terminal in another collocated network with enough power to cause an error in the data of the desired received signal. Although they operate on the same frequency band, the nature of Bluetooth and Wi-Fi signals are very different. Bluetooth has a narrow-band transmission of 1 MHz bandwidth which hops around pseudo-randomly over an 80 MHz band while Wi-Fi (using DSSS) has a broad, approximately 20 MHz, bandwidth that is constant in some region of the band. The interference phenomenon is apparent in Figure 11-25. Whenever there is a frequency and time coincidence of the transmission of one system and reception of the other, it's possible for an error to occur. Whether it does or not depends on the relative signal strengths of the desired and undesired signals. These in turn depend on the radiated power outputs of the transmitters and the distance between them and the receiver. When two terminals are very close (on the order of centimeters), interference may occur even when the transmitting frequency is outside the bandwidth of the affected receiver.

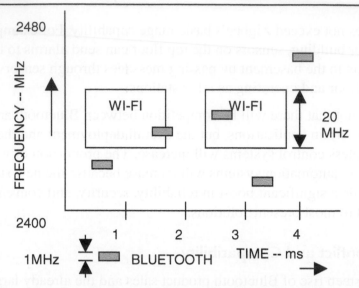

Figure 11-25: Wi-Fi and Bluetooth Spectrum Occupation

Bluetooth and Wi-Fi systems are not synchronous and interference between them has to be quantified statistically. We talk about the probability of a packet error of one system caused by the other system. The consequence of a packet error is that the packet will have to be retransmitted once or more until it is correctly received, which causes a delay in message throughput. Voice transmissions generally don't allow packet retransmission because throughput cannot be delayed, so interference results in a decrease in message quality.

Following are parameters that affect interference between Bluetooth and Wi-Fi:

- *Frequency and time overlap.* A collision occurs when the interferer transmits at the same time as the desired transmitter and is strong enough to cause a bit or symbol error in the received packet.

- *Packet length.* The longer the packet length of the Wi-Fi system, relative to a constant packet length and hop rate of Bluetooth, the longer the victim may be exposed to interference from one or more collisions and the greater the probability of a packet error.

- *Bit rate*. Generally, the higher the bit rate, the lower the receiver sensitivity and therefore the more susceptible the victim will be to packet error for given desired and interfering signal strengths. On the other hand, higher bit rates usually result in reduced packet length, with the opposite effect.

- *Use factor*. Obviously, the more often the interferer transmits, the higher the probability of packet error. When both communicating terminals of the interferer are in the interfering vicinity of the victim the use factor is higher than if the terminals are further apart and one of them does not have adequate strength to interfere with the victim.

- *Relative distances and powers*. The received power depends on the power of the transmitter and its distance. Generally, Wi-Fi systems use more power than Bluetooth, typically 20 mW compared to 1 mW. Bluetooth Class 1 systems may transmit up to 100 mW, but their output is controlled to have only enough power to give a required signal level at the receiving terminal.

- Signal-to-interference ratio of the victim receiver, SIR, for a specified symbol or frame error ratio.

- Type of modulation, and whether error-correction coding is used.

A general configuration for the location of Wi-Fi and interfering Bluetooth terminals is given in Figure 11-26. In this discussion, only transmissions from the access point to the mobile terminal are considered. We can get an idea of the vicinity around the Wi-Fi mobile terminal in which operating Bluetooth terminals will affect transmissions from the access point to the mobile terminal by examining the following parameters:

CI_{cc}, CI_{ac}—Ratio of signal carrier power to co-channel or adjacent channel interfering power for a given bit or packet error rate (probability).

P_{WF}, P_{BT}—Wi-Fi and Bluetooth radiated power outputs.

$PL = Kd^r$—Path loss which is a function of distance d between transmitting and receiving terminals, and the propagation exponent r. K is a constant.

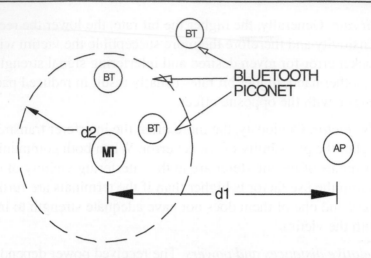

Figure 11-26: Importance of Relative Terminal Location

*d*1—Distance between Wi-Fi mobile terminal and access point.

*d*2—Radius of area around mobile terminal within which an interfering Bluetooth transmitter signal will increase the Wi-Fi bit error rate above a certain threshold.

PR_{WF}, PR_{BT},—Received powers from the access point and from the Bluetooth interfering transmitters.

*d*2 as a function of *d*1 is found as follows, using power in dBm:

1) $PR_{WF} = P_{WF} - 10\log(Kd1^r)$; $PR_{BT} = P_{BT} - 10\log(Kd2^r)$

2) $CI_{cc} = PR_{WF} - PR_{BT} = P_{WF} - 10\log(Kd1^r) - P_{BT} + 10\log(Kd2^r)$

3) $(CI_{cc} - P_{WF} + P_{BT})/10r = \log(d2/d1)$

4) $d2 = d1 \cdot 10^{(CI_{cc} - P_{WF} + P_{BT})/(10 \cdot r)}$ (11-1)

As an example, the interfering area radius *d*2 is now calculated from equation (11.1) using the following system parameters:

$CI_{cc} = 10$ dB, $P_{WF} = 13$ dBm, $P_{BT} = 0$ dBm, $r = 2$ (free space)

Solving equation (11-1): $d2 = d1 \times .71$

In this case, if a Wi-Fi terminal is located 15 meters from an access point, for example, all active Bluetooth devices within a distance of 10.6 meters from it have the potential of interfering. Only co-channel interference is considered. Adjacent channel interference, if significant, would increase packet error probability because many more Bluetooth hop channels would cause symbol errors. However, the adjacent channel CI_{ac} is on the order of 45 dB lower than CI_{cc} and would be noticed only when Bluetooth is several centimeters away from the Wi-Fi terminal.

The effect of an environment where path loss is greater than in free space can be seen by using an exponent $r = 3$. For the same Wi-Fi range of 15 meters, the radius of Bluetooth interference becomes 11.9 meters.

While equation (11-1) does give a useful insight into the range where Bluetooth devices are liable to deteriorate Wi-Fi performance, its development did involve simplifications. It considered that the signal-to-interference ratio that causes the error probability to exceed a threshold is constant for all wanted signal levels, which isn't necessarily so. It also implies a step relationship between signal-to-interference ratio and performance degradation, whereas the effect of changing interference level is continuous. The propagation law used in the development is also an approximation.

Methods for Improving Bluetooth and Wi-Fi Coexistence

By dynamically modifying one or more system operating parameters according to detected interference levels, coexistence between Bluetooth and Wi-Fi can be improved. Some of these methods are discussed below.

Power Control

Limiting transmitter power to the maximum required for a satisfactory level of performance will reduce interference to collocated networks. Power control is mandatory for Class 1 Bluetooth systems, where maximum power is 100 mW. The effect of the power on the interference radius is evident in equation (11-1). For example, in a Bluetooth piconet established between devices located over a spread of distances from the master, the master will use only the power level needed to communicate with each of the slaves in the network. Lack of power control would mean that all devices would communicate at maximum power and the collocated Wi-Fi system would be exposed to a high rate of interfering Bluetooth packets.

Adaptive Frequency Hopping

Wi-Fi and Bluetooth share approximately 25 percent of the total Bluetooth hop-span of 80 MHz. Probably the most effective way to avoid interference between the two systems is to restrict Bluetooth hopping to the frequency range not used by Wi-Fi. When there is no coordination or cooperation between collocated networks, the Bluetooth piconet master would have to sense the presence of Wi-Fi transmissions and modify the frequency-hopping scheme of the network accordingly. A serious obstacle to this method was lifted by a change to the FCC regulations governing spread-spectrum transmissions in the 2.4 GHz band. Previously, frequency hopping devices were committed to hopping over at least 75 pseudo-randomly selected hop channels. In August 2002, paragraph 15.247, according to which Bluetooth and Wi-Fi devices are regulated, was changed to allow a minimum of 15 nonoverlapping channels in the 2400 to 2483 MHz band. In addition, the regulation allows employing intelligent hopping techniques, when less than 75 hopping frequencies are used, to avoid interference with other transmissions, and also suppression of transmission on an occupied channel provided that there are a minimum of 15 hops. The Bluetooth specification is due to be modified to take advantage of the adaptive frequency hopping method of avoiding interference.

There are situations where adaptive frequency hopping may not be effective or may have a negative effect. When two or more adjacent Wi-Fi networks are operating concurrently, they will utilize different 22 MHz sections of the 2.4 GHz band—three nonoverlapping Wi-Fi channels are possible. In this case, Bluetooth may not be able to avoid collisions while using a minimum of 15 hop frequencies. In addition, if there are several Bluetooth piconets in the same area, collisions among themselves will be much more frequent than when the full 79 channel hopping sequences are used.

Packet fragmentation

The two interference-avoiding methods described above are applicable primarily for action by the Bluetooth network. One method that the Wi-Fi network can employ to improve throughput is packet fragmentation. By fragmenting data packets and sending more, but shorter transmission frames, each transmission will have a lower probability of collision with a Bluetooth packet. Although reducing frame size increases the percentage of overhead bits in the transmission, when interference is heavy the overall effect may be higher throughput than if fragmentation was not used. Increasing bit rate for a constant packet length will also result in a shorter transmitted frame and less exposure to interference.

The methods mentioned above for reducing interference presume no coordination between the two different types of collocated wireless networks. However, devices are now being produced, in laptop and notebook computers for example, that include both Wi-Fi and Bluetooth, sometimes even in the same chipset. In this case collaboration is possible in the device software to prevent inter-network collisions.

11.5 Ultra-wideband Technology

Ultra-wideband (UWB) technology is based on transmission of very narrow electromagnetic pulses at a low repetition rate. The result is a radio spectrum that is spread over a very wide bandwidth—much wider than the bandwidth used in the spread-spectrum systems previously discussed. Ultra-wideband transmissions are virtually undetectable by ordinary radio receivers and therefore can exist concurrently with existing wireless communications without demanding additional spectrum or exclusive frequency bands.

These are some of the advantages cited for ultra-wideband technology:

- Very low spectral density—Very low probability of interference with other radio signals over its wide bandwidth,

- High immunity to interference from other radio systems,

- Low probability of interception/detection by other than the desired communication link terminals,

- High multipath immunity,

- Many high data rate ultra-wideband channels can operate concurrently,

- Fine range-resolution capability,

- Relatively simple, low-cost construction, based on nearly all digital architectures.

Transmission and reception methods are unique, and are described briefly below.

Differing from conventional radio communication systems, which use up conversion and down conversion to pass information signals between baseband and bandpass frequency channels where wireless propagation occurs, UWB signal generation and detection use baseband techniques. An example of a UWB "carrier" is a Gaussian monopulse, shown in Figure 11-27 (Reference 43). Its power spectrum is shown in Figure 11-28. If the time scale in Figure 11-27 is in nanoseconds, then the width of the pulse is 0.5 nanoseconds and the 3 dB bandwidth of the power spectrum is approximately 3.2 GHz with maximum power density at 2 GHz.

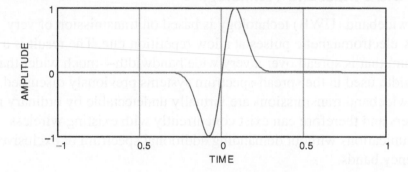

Figure 11-27: UWB Monopulse

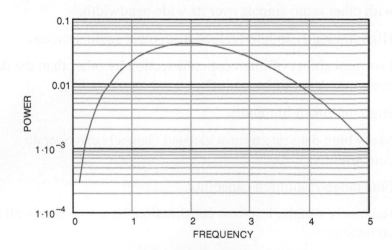

Figure 11-28: Spectrum of Monopulse

In order to pass information over a UWB communication link, trains of pulses must be transmitted with some characteristic of a pulse or group of pulses varied in order to distinguish between "0" and "1." The time between consecutive pulses should be determined in a pseudo-random manner in order to smooth the energy spikes in the frequency spectrum. Reception of the transmitted pulse train is done by correlating the received signal with a similar sequence of pulses generated in the receiver. A large number of communication links can be maintained simultaneously and independently by using different pseudo-random sequences for each link.

A pulse similar to that of Figure 11-27 can be generated by applying an impulse, or perhaps more conveniently a step-voltage or current, to a linear band limited network. Figure 11-29 is a simulation of a sequence of UWB pulses created by stimulating a bandpass filter with a pseudo-

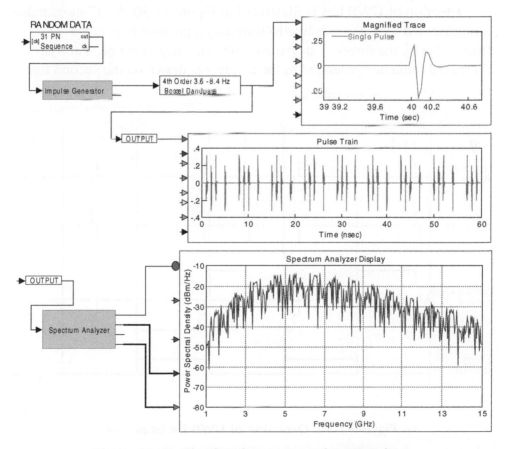

Figure 11-29: Simulated Sequence of UWB Pulses

randomly spaced sequence of impulses. The figure also shows the power spectrum of that sequence. The network that creates the individual UWB pulses includes the transmitter antenna, the propagation channel, and the receiving antenna, whose characteristics, in terms of impulse response or amplitude and phase vs. frequency must be known and accounted for in designing the system.

There are several ways of representing a UWB pulse as "1" or "0." One method is to advance or retard the transmitted pulse with respect to the expected time of arrival of the pulse in the receiver according to the agreed pseudo-random time sequence. Another method is to send the pulse with or without inversion. In both cases the correlation of the received pulse with a "template" pulse generated in the receiver will result in a different polarity, depending on whether a "1" or a "0" was transmitted.

Detection of UWB bits is illustrated in Figure 11-30. A "1" monopulse is represented by a negative line followed by a positive line, and a "0" monopulse by the inverse—a positive line and a negative line. The synchronized sequence generated in the receiver is drawn on the second line

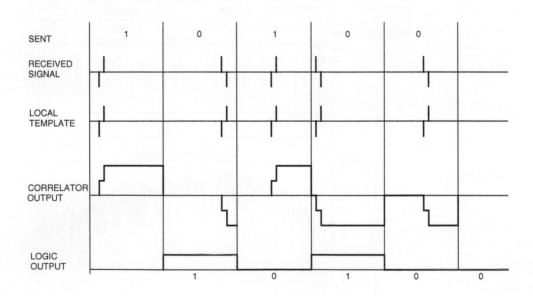

Figure 11-30: Detection of UWB Bit Sequence

and below it the result of the correlation operation $\int f(t) \cdot g(t) dt$ where $f(t)$ is the received signal and $g(t)$ is the locally generated sequence. By sampling this output at the end of each bit period and then resetting the correlator, the transmitted sequence is reconstructed in the receiver.

As mentioned above, an individual bit can be represented by more than one sequential monopulse. Doing so increases the processing gain by the number of monopulses per bit. Processing gain is also an inverse function of the pulse duty cycle. This is because, for constant average power, the power in the pulses contributing to each bit must be raised by (1/*duty-cycle*). By gating out the noise except during the interval of the expected incoming pulse, the signal-to-noise ratio will only be a function of the power in the pulse, regardless of the duty cycle. An example may make the explanation clearer. Let's say we are sending data a rate of 10 Mbps. A UWB pulse in the transmitter is 200 picoseconds wide; 20 pulses represent one bit. The time between bits is $1/(10 * 10^6) = 100$ nanoseconds. So the time between pulses is (100 ns)/20 = 5 ns. The duty cycle is (200 ps)/(5 ns) = 25. Now the processing gain attributed to the number of UWB pulses per bit is $10\log(20) = 13$ dB. That due to the duty cycle is $10\log(25) = 14$ dB. Total processing gain is $13 + 14 = 27$ dB.

A simplified block diagram of a UWB system is shown in Figure 11-31. A key to the generation of UWB pulses is the ability to create short impulse or step functions with rise times on the order of tens or at the most hundreds of picoseconds, and to detect the UWB pulses that result from their application. High speed integrated circuits can be employed or special circuit elements, such as tunnel diodes or step recovery diodes, can be incorporated.

The main FCC regulations pertaining to ultra-wideband transmissions are described in Chapter 9. Conditions for using UWB in Europe are also being considered by the European Union. Due to the fear of interference with vital wireless services in the wide bandwidths covered by UWB radiation, spectral density limits allowed by the FCC and presumably to be permitted in the European Union are relatively low, on the order of spurious radiation limits for conventional unlicensed transmissions. However, as we have seen, high-processing gains can be achieved with UWB and communication ranges on the order of tens or hundreds of meters at high-data rates can be expected. Also, the FCC has indicated its intention to monitor the effects of UWB transmissions on other services, once equipment has been put into service in significant quantities, and the agency may be expected to modify or make its limits more lenient if interference is found not to be a problem. In any case, the unique characteristics of UWB are attractive enough to make this technology an important part of the offerings for short-range wireless communication in the years to come.

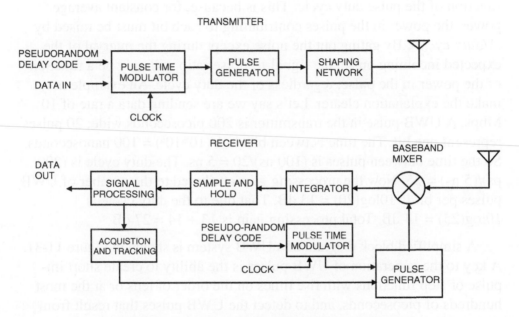

Figure 11-31: UWB simplified block diagram

11.6 Summary

The annually increasing volumes for Bluetooth and Wi-Fi products, stimulated in a large part by the acceptance of industrial standards by the major manufacturers, are causing prices to fall on complex integrated circuits as well as the basic RF components. This trend will open the way for the use of these parts in other short-range applications such as security and medical call systems. Use of sophisticated and proven two-way hardware and link protocols for these and other technically "low end" applications will open them up to much higher usage than they now command. A basic impediment to wireless use will still remain, however, and that is the problem of battery replacement. Reduced voltage and power consumption for integrated circuits will help, as will sophisticated wake-up protocols as are already built-in to Bluetooth. Range limitations may have to be dealt with by a greater use of repeaters than common in today's systems. Another area where advancements are affecting short-range radio is antennas. Both the use of higher frequencies and new designs are reducing antenna size and eliminating a visual reminder of the difference between wired and wireless devices.

The unconventional ultra-wideband technology, since its approval by the FCC, is opening up new civilian applications for short-range wireless, notably in the areas of distance measurement, concealed object location, and high precision positioning systems. Because of its high-interference immunity, and its property of not causing interference, it may successfully compete with and complement other technologies used for short-range radio applications such as personal communications systems, security sensors, and RFID tags.

In summary, advances in short-range radio communication developments in one area feeds its expansion in other areas. Overall, short-range radio will continue to play a major part in the ongoing communication revolution.

Abbreviations

2FSK	Frequency shift keying with two frequencies
4FSK	Frequency shift keying with four frequencies
ADC	Analog-to-digital converter
AGC	Automatic gain control
AM	Amplitude modulation
AP	Access point
ARQ	Automatic repeat request or automatic repeat query
ASH	Amplifier sequenced hybrid (receiver)
ASIC	Application specific integrated circuit
ASK	Amplitude shift keying
BER	Bit error rate
CCK	Complementary code keying
CDMA	Code division multiple access
CDROM	Compact disk read-only memory
CENELEC	European Committee for Electrotechnical Standardization
CEPT	European Conference of Postal and Telecommunication Administrations
CSMA/CA	Carrier sense multiple access with collision avoidance
DAC	Digital-to-analog converter
DDS	Direct digital synthesizer
DPSK	Differential phase shift keying
DSSS	Direct sequence spread spectrum
EIRP or e.i.r.p.	Equivalent isotropic radiated power
EMC	Electromagnetic compatibility
ERP or e.r.p.	Effective radiated power (from dipole antenna)
ETSI	European Telecommunication Standards Institute
EU	European Union
FCC	Federal Communications Commission
FEC	Forward error correction
FHSS	Frequency-hopping spread spectrum

FM	Frequency modulation
FRS	Family radio service
FSK	Frequency shift keying
GMSK	Gaussian minimum shift keying
GSM	Global system for mobile communications
HIPERLAN	High performance local area network
IC	Integrated circuit
IF	Intermediate frequency
ISM	Industrial, scientific, medical (frequency bands)
LC	Inductive-capacitive or Inductor-capacitor
LNA	Low noise amplifier
LPRS	Low power radio service
MDS	Minimum discernable signal
OFDM	Orthogonal frequency division multiplex
PD	Phase detector
PLL	Phase-locked loop
Q	Quality factor
QAM	Quadrature amplitude modulation
RFID	Radio frequency identification
RMS	Root mean square
RSSI	Receive signal strength indicator
R&TTE	Radio and telecommunications terminal equipment
SAW	Surface acoustic wave (device)
SRD	Short-range device
SWAP	Shared wireless access protocol
TDMA	Time division multiple access
TRF	Tuned radio frequency (receiver)
UHF	Ultra high frequency (300 to 3000 MHz)
U-NII	Unlicensed national information infrastructure
UWB	Ultra-wideband
VCO	Voltage controlled oscillator
VFO	Variable frequency oscillator
VHF	Very high frequency (30 to 300 MHz)
WBFM	Wide-band frequency modulation
WLAN	Wireless local area network
WPAN	Wireless personal area network

References and Bibliography

References supplementary to text are designated by chapter number in brackets [ ].

1. "A Saw Stabilized FSK Oscillator," RF Monolithics. [5]

2. "High Performance Low Power FM IF System," Product Specification NE/SA604A, Philips, 1990.

3. "Low-Cost Integrated Solution for Analog Cellular RF Block," Application Note TA0009, RF Micro-Devices.

4. "Reviewing Key Areas When Designing With the NE605," Application Note An1994, Philips, 1985. [8]

5. ANSI/IEEE Std 802.11, 1999 Edition, Part 11: Wireless LAN Medium Access Control (MAC) and Physical Layer (PHY) Specifications.

6. Anthes, John, "OOK, ASK and FSK Modulation in the Presence of an Interfering Signal," Application Note, RF Monolithics, Dallas, Texas. [4]

7. Application Note AN-282A, "Systemizing RF Power Amplifier Design," Motorola

8. Bensky, Alan, "Range Estimation for Short-range Event Transmission Systems," RF Design Magazine, November 2002, p. 30.

9. Bluetooth Special Interest Group, Specification of the Bluetooth System, Version 1.1, February 2001.

10. Buchman, Isador, "Choosing a Lasting Battery," Radio Resource International, Quarter 4 1999, p. 21. [7]

11. Carlson, Bruce A., *Communication Systems: An Introduction to Signals and Noise in Electrical Communication*, McGraw-Hill Book Company, New York, 1968.

12. Chipal Chou, Welman Si, Jasen Chen, "Reviewing Fractional-N Frequency Synthesis," Microwaves & RF, June 1999, p. 53. [5]

13. Dacus, Farron L., "Understanding Regulations for Short-range Radios," Microwaves & RF, October 2001, p. 79. [9]

14. Dacus, Farron L., Van Niekerk, Jan, and Bible, Steven, "Constructing Circuits for Short-range Radios, Microwaves & RF, February 2002, p. 59.

15. Dacus, Farron L., Van Niekerk, Jan, and Bible, Steven, "Introducing Loop Antennas for Short-range Radios," Microwaves & RF, July 2002, p. 80. [3]

16. Dacus, Farron L., Van Niekerk, Jan, and Bible, Steven, "Tracking Phase Noise in Short-range Radios," Microwaves & RF, March 2002, p. 57. [7]

17. Dixon, Robert C., *Spread Spectrum Systems*, John Wiley & Sons, New York, 1984. [4]

18. Drabowitch, S., Papiernik, A., Griffiths, H., Encinas, J. *Modern Antennas*, Chapman & Hall, London, 1998. [3]

19. Draft P802.15.4/D18, February-2003, Draft Standard for Part 15.4: Wireless Medium Access Control (MAC) and Physical Layer (PHY) specifications for Low Rate Wireless Personal Area Networks (LR-WPANs) (Zigbee).

20. ETSI TR 101 031 V2.2.1 (1999-01) Technical Report, HIPERLAN/2 Requirements and architectures for wireless broadband access.

21. ETSI TR 101 683 V1.1.1 (2000-02) Technical Report, HIPERLAN/2 System Overview.

22. ETSI TS 101 475 V1.3.1 (2001-12) Technical Report, HIPERLAN/2 Physical (PHY) Layer.

23. Fu, T. L., "Optimize the Performance of Pager Antennas," Microwaves & RF, August 1994, p. 141.

24. Fujimoto, K., Henderson, A., Hirasawa, K., and James, J. R., *Small Antennas*, Research Studies Press, Ltd, England, 1987. [3]

25. Gallager, Robert G., *Information Theory and Reliable Communication*, John Wiley and Sons, Inc., New York, 1968.

26. Gibson, Jerry D., Editor-in-Chief, *The Mobile Communications Handbook*, CRC Press, Inc., 1996.

27. Hall, Barclay, and Hewitt, *Propagation of Radio Waves*, IEE, 1996.

28. Hall, Gerald, Editor, *The ARRL Antenna Book,* The American Radio Relay League, Newington, CT, USA, 1991.

29. Hall, Gerald, Editor, *The ARRL Antenna Book,* The American Radio Relay League, Newington , CT, U.S.A. 1991.

30. Hayward, Wes and DeMaw, Doug, *Solid State Design For the Radio Amateur,* American Radio Relay League, Inc. Newington, CT, 1986.

31. Heftman, Gene, "Cellular IC's Move Toward 3G Wireless – Gingerly," Microwaves & RF, February 99, p. 31. [6]

32. IEEE Std 802.11a-1999, High-speed Physical Layer in the 5 GHz Band.

33. IEEE Std 802.11b-1999, Higher-Speed Physical Layer Extension in the 2.4 GHz Band.

34. IEEE Std 802.15.1™-2002, Part 15.1: Wireless Medium Access Control (MAC) and Physical Layer (PHY) Specifications for Wireless Personal Area Networks (WPANs) (Bluetooth).

35. Jacobs, Sol, "How to Choose a Primary Battery," Tadiran Battery Division, Port Washington, NY. [7]

36. Jasik, Henry, Editor, *Antenna Engineering Handbook*, McGraw-Hill, NY, 1961. [3]

37. Kleinschmidt, Kirk A., Editor, *The ARRL Handbook for the Radio Amateur,* American Radio Relay League, Newington, CT, USA, 1990.

38. Larson, Lawrence E., *RF and Microwave Circuit Design for Wireless Communications,* Artech House, Inc., Norwood, MA, 1997.

39. Lo, Y.T., Lee, S.W., *Antenna Handbook, Theory, Applications and Design*, Van Nostrand Reinhold Company, Inc., New York, 1988.

40. Milligan, Thomas A., *Modern Antenna Design*, McGraw-Hill, NY, 1985 [3]

41. Motorola Specification and Application Notes for MC4344/MC4044 Phase-Frequency Detector.

42. Papoulis, Athanasios, *Probability, Random Variables, and Stochastic Processes,* McGraw-Hill, New York, 1965.

43. Petroff, Alan and Withington, Paul, "Time Modulated Ultra-Wideband (TM-UWB) Overview," Presented at Wireless Symposium/Portable by Design, Feb 25, 2000, San Jose, California (http://www.time-domain.com). [11]

44. Rambabu, K., Ramesh, M., and Kalghatgi, A. T., "Antenna Design Offers Small Size With Wide Bandwidth," Microwaves & RF, September 1997, p. 105.

45. Rappaport, Theodore S., *Wireless Communications, Principles and Practice*, Prentice Hall, Upper Saddle River, NJ, 1996. [2]

46. *Reference Data for Radio Engineers*, Fourth Edition, International Telephone and Telegraph Corporation, New York, 1956.

47. Reza, Fazlollah M., *An Introduction To Information Theory*, McGraw-Hill Book Company, Inc. New York, 1961.

48. RFM Product Data Book, 1997.

49. Robert, B., Razban, T., and Papiernik, A., "Compact Patch Antenna Integrates Monolithic Amp," Microwaves & RF, March 1995, p. 115.

50. Rohde, Ulrich L., "Fractional-N Methods Tune Base-Station Synthesizer," Microwaves & RF, April 1998, p. 151. [5]

51. Sklar, Bernard, "Digital Communications, Fundamentals and Applications, Prentice Hall, Upper Saddle River, NJ, 2001.

52. Smith Greg, "Solid Polymer Lithium-ion Batteries Provide a Good Fit for Wireless Designs," Wireless Systems Design, August 1998, p. 41. [7]

53. Smith, Kent, "Antennas for Low Power Applications," Application Note AN36, RF Monolithics, Dallas, Texas.

54. Spix, George J., "Maxwell's Electromagnetic Field Equations," unpublished tutorial, copyright 1995 (http://www.connectos.com/spix/rd/gj/nme/maxwell.htm). [2]

55. Stutzman & Thiele, *Antenna Theory and Design*, John Wiley & Sons, Inc., 1998. [3]

56. Taylor, Larry, "HIPERLAN Type 1 Technology Overview," TTP Communications LTD, 1999.

57. Terman, Frederick E., *Electronic & Radio Engineering*, McGraw Hill, New York, 1955.

58. Van Niekerk, Jan, Dacus, Farron L., and Bible, Steven, "Matching Loop Antennas to Short-range Radios," Microwaves & RF, August 2002, p. 72. [3]

59. Van Valkenburg, Editor-in-Chief, *Reference Data for Engineers*, Eighth Edition, SAMS, Prentice Hall, Carmel, Indiana, 1993.

60. Vear, Tim, "Selection and Operation of Wireless Microphone Systems," Shure Brothers Inc. 1998. [4]

61. Weeks, W. L., *Antenna Engineering*, McGraw Hill, NY, 1968. [3]

62. Wong, Michael, and Placer, Maria Lourdas, "Designing Receiver Front Ends From System Specifications," RF Design, April 1998, p. 44. [7]

63. Yestrebsky, Tom, "MICRF001 Antenna Design Tutorial," Application Note 23, Micrel, Inc. San Jose, California. [3]

Web Sites

Bluetooth Special Interest Group (SIG) – www.bluetooth.com/

ETSI – www.etsi.fr/

European Radiocommunication Organization (ERO) – www.ero.dk/

Federal Communications Commission – www.fcc.gov/oet/

HIPERLAN/2 Global Forum – www.hiperlan2.com

Institute of Electrical and Electronic Engineers – www.ieee.org

Japanese Telecommunications Ministry – www.tele.soumu.go.jp/e/

Low Power Radio Association – www.lpra.org/

Microwaves & RF Magazine – www.planetee.com

Radiocommunications Agency, UK – www.radio.gov.uk/

RF Design Magazine – www.rfdesign.com

Shure – www.shure.com/

Ultra-wideband Working Group – www.uwb.org/

Wi-Fi Alliance – www.weca.net

Zigbee Alliance – www.zigbee.org

Index

TERM

This Agreement will remain in effect until terminated pursuant to the terms of this Agreement. You may terminate this Agreement at any time by removing from Your system and destroying the CD-ROM Product. Unauthorized copying of the CD-ROM Product, including without limitation, the Proprietary Material and documentation, or otherwise failing to comply with the terms and conditions of this Agreement shall result in automatic termination of this license and will make available to Elsevier Science legal remedies. Upon termination of this Agreement, the license granted herein will terminate and You must immediately destroy the CD-ROM Product and accompanying documentation. All provisions relating to proprietary rights shall survive termination of this Agreement.

LIMITED WARRANTY AND LIMITATION OF LIABILITY

NEITHER ELSEVIER SCIENCE NOR ITS LICENSORS REPRESENT OR WARRANT THAT THE INFORMATION CONTAINED IN THE PROPRIETARY MATERIALS IS COMPLETE OR FREE FROM ERROR, AND NEITHER ASSUMES, AND BOTH EXPRESSLY DISCLAIM, ANY LIABILITY TO ANY PERSON FOR ANY LOSS OR DAMAGE CAUSED BY ERRORS OR OMISSIONS IN THE PROPRIETARY MATERIAL, WHETHER SUCH ERRORS OR OMISSIONS RESULT FROM NEGLIGENCE, ACCIDENT, OR ANY OTHER CAUSE. IN ADDITION, NEITHER ELSEVIER SCIENCE NOR ITS LICENSORS MAKE ANY REPRESENTATIONS OR WARRANTIES, EITHER EXPRESS OR IMPLIED, REGARDING THE PERFORMANCE OF YOUR NETWORK OR COMPUTER SYSTEM WHEN USED IN CONJUNCTION WITH THE CD-ROM PRODUCT.

If this CD-ROM Product is defective, Elsevier Science will replace it at no charge if the defective CD-ROM Product is returned to Elsevier Science within sixty (60) days (or the greatest period allowable by applicable law) from the date of shipment.

Elsevier Science warrants that the software embodied in this CD-ROM Product will perform in substantial compliance with the documentation supplied in this CD-ROM Product. If You report significant defect in performance in writing to Elsevier Science, and Elsevier Science is not able to correct same within sixty (60) days after its receipt of Your notification, You may return this CD-ROM Product, including all copies and documentation, to Elsevier Science and Elsevier Science will refund Your money.

YOU UNDERSTAND THAT, EXCEPT FOR THE 60-DAY LIMITED WARRANTY RECITED ABOVE, ELSEVIER SCIENCE, ITS AFFILIATES, LICENSORS, SUPPLIERS AND AGENTS, MAKE NO WARRANTIES, EXPRESSED OR IMPLIED, WITH RESPECT TO THE CD-ROM PRODUCT, INCLUDING, WITHOUT LIMITATION THE PROPRIETARY MATERIAL, AN SPECIFICALLY DISCLAIM ANY WARRANTY OF MERCHANTABILITY OR FITNESS FOR A PARTICULAR PURPOSE.

If the information provided on this CD-ROM contains medical or health sciences information, it is intended for professional use within the medical field. Information about medical treatment or drug dosages is intended strictly for professional use, and because of rapid advances in the medical sciences, independent verification f diagnosis and drug dosages should be made.

IN NO EVENT WILL ELSEVIER SCIENCE, ITS AFFILIATES, LICENSORS, SUPPLIERS OR AGENTS, BE LIABLE TO YOU FOR ANY DAMAGES, INCLUDING, WITHOUT LIMITATION, ANY LOST PROFITS, LOST SAVINGS OR OTHER INCIDENTAL OR CONSEQUENTIAL DAMAGES, ARISING OUT OF YOUR USE OR INABILITY TO USE THE CD-ROM PRODUCT REGARDLESS OF WHETHER SUCH DAMAGES ARE FORESEEABLE OR WHETHER SUCH DAMAGES ARE DEEMED TO RESULT FROM THE FAILURE OR INADEQUACY OF ANY EXCLUSIVE OR OTHER REMEDY.

U.S. GOVERNMENT RESTRICTED RIGHTS

The CD-ROM Product and documentation are provided with restricted rights. Use, duplication or disclosure by the U.S. Government is subject to restrictions as set forth in subparagraphs (a) through (d) of the Commercial Computer Restricted Rights clause at FAR 52.22719 or in subparagraph (c)(1)(ii) of the Rights in Technical Data and Computer Software clause at DFARS 252.2277013, or at 252.2117015, as applicable. Contractor/Manufacturer is Elsevier Science Inc., 655 Avenue of the Americas, New York, NY 10010-5107 USA.

GOVERNING LAW

This Agreement shall be governed by the laws of the State of New York, USA. In any dispute arising out of this Agreement, you and Elsevier Science each consent to the exclusive personal jurisdiction and venue in the state and federal courts within New York County, New York, USA.

Printed and bound by CPI Group (UK) Ltd, Croydon, CR0 4YY

03/10/2024

01040331-0002